Mathematical Reminiscences

Howard Eves

Mathematical Reminiscences

Howard W. Eves

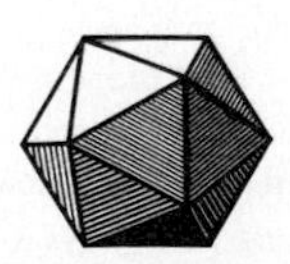

Published and Distributed by
The Mathematical Association of America

Library of Congress Catalog Card Number 2001090643

ISBN 0-88385-535-6

Printed in the United States of America

Current printing (last digit):
10 9 8 7 6 5 4 3 2 1

The Spectrum Series of the Mathematical Association of America was so named to reflect its purpose: to publish a broad range of books including biographies, accessible expositions of old or new mathematical ideas, reprints and revisions of excellent out-of-print books, popular works, and other monographs of high interest that will appeal to a broad range of readers, including students and teachers of mathematics, mathematical amateurs, and researchers.

All the Math That's Fit to Print, by Keith Devlin

Circles: A Mathematical View, by Dan Pedoe

Complex Numbers and Geometry, by Liang-shin Hahn

Cryptology, by Albrecht Beutelspacher

Five Hundred Mathematical Challenges, Edward J. Barbeau, Murray S. Klamkin, and William O. J. Moser

From Zero to Infinity, by Constance Reid

The Golden Section, by Hans Walser. Translated from the original German by Peter Hilton, with the assistance of Jean Pedersen.

I Want to Be a Mathematician, by Paul R. Halmos

Journey into Geometries, by Marta Sved

JULIA: a life in mathematics, by Constance Reid

The Lighter Side of Mathematics: Proceedings of the Eugène Strens Memorial Conference on Recreational Mathematics & Its History, edited by Richard K. Guy and Robert E. Woodrow

Lure of the Integers, by Joe Roberts

Magic Tricks, Card Shuffling, and Dynamic Computer Memories: The Mathematics of the Perfect Shuffle, by S. Brent Morris

The Math Chat Book, by Frank Morgan

Mathematical Carnival, by Martin Gardner

Mathematical Circus, by Martin Gardner

Mathematical Cranks, by Underwood Dudley

Mathematical Fallacies, Flaws, and Flimflam, by Edward J. Barbeau

Mathematical Magic Show, by Martin Gardner

Mathematical Reminiscences, by Howard Eves

Mathematics: Queen and Servant of Science, by E.T. Bell

Memorabilia Mathematica, by Robert Edouard Moritz

New Mathematical Diversions, by Martin Gardner

Non-Euclidean Geometry, by H. S. M. Coxeter

Numerical Methods That Work, by Forman Acton

Numerology or What Pythagoras Wrought, by Underwood Dudley

Out of the Mouths of Mathematicians, by Rosemary Schmalz

Penrose Tiles to Trapdoor Ciphers ... and the Return of Dr. Matrix, by Martin Gardner

Polyominoes, by George Martin

Power Play, by Edward J. Barbeau

The Random Walks of George Pólya, by Gerald L. Alexanderson

The Search for E.T. Bell, also known as John Taine, by Constance Reid

Shaping Space, edited by Marjorie Senechal and George Fleck

Student Research Projects in Calculus, by Marcus Cohen, Arthur Knoebel, Edward D. Gaughan, Douglas S. Kurtz, and David Pengelley

Symmetry, by Hans Walser. Translated from the original German by Peter Hilton, with the assistance of Jean Pedersen.

The Trisectors, by Underwood Dudley

Twenty Years Before the Blackboard, by Michael Stueben with Diane Sandford

The Words of Mathematics, by Steven Schwartzman

MAA Service Center
P.O. Box 91112
Washington, DC 20090-1112
800-331-1622   FAX 301-206-9789

Foreword

I suppose it is natural late in one's life to look back and enjoy again some of the more precious moments of earlier life. A number of my friends have urged me to record a collection of these earlier moments. It didn't take much reflection to realize that, if I were to keep the treatment within reasonable bounds, I must limit myself by some rule of selection. I decided to concentrate chiefly on those moments more or less connected either with mathematics or with teaching, two pursuits that have occupied such a large part of my life.

And so, with deep apologies for the naiveté of my presentations, and with the kind forbearance of my understanding readers, I here offer the following potpourri of fifty nostalgic tales, thoughts, and escapades. They are presented in no particular logical order, but rather as they occurred to me. Some of the tales required only a few lines while others needed a few pages.

Acknowledgements

I thank *The Fibonacci Quarterly* for permitting me to repeat my article "Hail to Thee, Blithe Spirit!" (concerning Vern Hoggatt), and the PWS Publishing Company for permitting me to adapt some of the personal items that have appeared in my *Circle* books.* The *Circle* books contain many other items concerning mathematics and the teaching of mathematics. Versions of four of these tales have appeared in *In Eves' Circles,* Notes No. 34 of the MAA, and some others have been narrated at various mathematics gatherings.

The author deeply appreciates the fine assistance of Professor Jerry Alexanderson in the preparation of the final manuscript of the present work.

**In Mathematical Circles, Mathematical Circles Revisited, Mathematical Circles Squared, Mathematical Circles Adieu, Return to Mathematical Circles.*

Contents

The Mystery of the Four-Leaf Clovers

It was in 1955 that the Hafner Publishing Company of New York City brought out G. W. Dunnington's scholarly *Carl Friedrich Gauss, Titan of Science: A Study of His Life and Work*. I was teaching at the University of Maine at the time and, as I was then in the process of building up a history of mathematics library, I ordered a copy of the book. I keenly looked forward to the book's arrival, and was delighted a couple of weeks later to find it on my office desk, amongst an assortment of other mail, delivered there for me while I was in class. Since I had a free period, I sat down and began listlessly paging through the volume. Imagine my astonishment when I got to the facing pages 100 and 101 to find there four neatly pressed four-leaf clovers. What a mysterious surprise! For some days I pondered on the mystery and told a number of my colleagues about it. Many wild and unlikely conjectures were offered, like: perhaps the book company does this with all their books, or perhaps only with the Dunnington volume, or maybe only for new customers, and so on. Finally my friend Professor Wootton decided to order a copy of the book for himself just to see if he too would receive some pressed four-leaf clovers. In time his copy arrived. It contained no four-leaf clovers.

And so the mystery remained until, after the passage of a couple of months, I received the following letter:

Dear Howard:

By now you must have found some four-leaf clovers inserted in a book you ordered from Hafner Publishing Company, and I bet you've been wondering how they got there. Here is the explanation. My wife Betty works in Hafner's shipping department. Some time ago she came home and said, "Ted, guess to whom I will mail off my first book tomorrow." I told her I had no idea. She replied, "To that crazy [her word not mine] rambling friend of yours, Howard Eves, who is now a professor at the University of Maine."

The name awoke wonderful memories and a thought immediately occurred to me. I went to the place where I keep some treasured mementos and selected four pressed four-leaf clovers from a collection of a dozen that I had there. I said to Betty, "Before you send off the book, place these clovers between pages 100 and 101 of the book, and we'll let Howard wonder how they got there."

Does this ring a bell? Recall our last ramble together of 20-some years ago, up the Jersey side of the Hudson River, where we searched in vain for Professor Kasner's hiding place and you performed an interesting mathematical experiment with a parabolic template. Recall our return along the top of the cliffs and how we each collected a dozen four-leaf clovers. There, now you know.

I am working as a researcher for an electronics firm in Newark, New Jersey, and about the only rambling I now do is in the firm's lab. Howard, I often think of you and talk about you. I shall never forget our friendship and our wonderful rambles together.

All best wishes,

Ted

Ted's letter certainly did ring a bell. It brought back to my mind a beautifully clear late-summer day of years ago. Ted and I were two young fellows each home from our second year in college, he from the Newark College of Engineering and I from the University of Virginia. Continuing a pattern set some years

My wife, Diane, and I at the Methodist Church, 1954, Orono, Maine

ago we had taken a number of rambles together during the summer, and Ted proposed we take one more before the summer vacation ended. He asked me to choose the place of the ramble.

Now it was our custom to choose places that had some (admittedly perhaps often very tenuous) connection with some important real or literary figure. Thus we once crossed the Tappan Zee and explored Tarrytown, ostensibly looking for the walnut grove in which Brom Bones hid when he flung a pumpkin at Ichabod Crane. Another time we ascended Slide Mountain in the Catskills up the slide itself, since it was this way the naturalist John Burroughs first climbed the mountain. Again, we once went up to Concord and Walden Pond in Massachusetts to stand on various spots described by Thoreau in his famous book and in his journals. Well, I had been reading about the lovable and eccentric Dr. Edward Kasner, Adrian Professor of Mathematics at Columbia University, who enjoyed nonfatiguing rambles. One of his favorite walks was up the Jersey side of the Hudson River at the base of the great basaltic cliffs known as the Palisades. Since he often took this walk he decided to carry only the perishable ingredients for his lunch, such as his tea and sugar, and to hide the teapot and other paraphernalia that he required in a box somewhere along the way. So I suggested that we follow in Professor Kasner's footsteps up the Hudson River and try to discover his hiding place.

We would, I said, take a bus from Paterson, New Jersey, where we were living, to the Jersey end of the George Washington suspension bridge linking New Jersey with New York across the Hudson River. There we would disembark and proceed on foot on the path following the base of the Palisades. This path continues for many miles up the Hudson River and is as flat and even as a sidewalk—a very suitable path for Professor Kasner, who was a stout man and had a great aversion to physical exertion. He used to boast that each summer when he vacationed in Belgium he would organize a mountain-climbing expedition to the highest peak in the country. When asked the elevation of the peak, he would say, "Twelve feet above sea level."

To Ted's mystification, before starting our ramble I put in my side pocket a carefully constructed parabola-shaped template cut from a piece of thin cardboard. Very shortly I shall explain the purpose of the template.

And so, on a bright Saturday morning, we set off on our jaunt, starting about 7:30 A.M. from the Jersey end of the great George Washington bridge. The Hudson at this point is a truly picturesque place, with the sparkling water, the boating and shipping on the river, and the buildings of upper New York City on the farther shore. We proceeded about a mile and a half up the river when we paused and looked back down the river, where we could see the George Washington suspension bridge of our starting point. I produced my parabolic template and launched into a short mathematical lecture.

You see, one of my early loves in mathematics was the field of differential equations. I had bought a couple of the then current textbooks on the subject and had started to teach myself the basics of this field of study when I was in my freshman year at college. There are many striking applications of differential equations, and one that particularly appealed to me concerned a cable hanging between two supports at the same height about the ground.

Clearly the curve assumed by a hanging cable depends upon how the weight of the cable is distributed along its length. It is well-known that if the weight is uniformly distributed along the cable (that is, so that the weights of arcs of the cable are proportional to the lengths of those arcs) then the cable hangs in a catenary curve. Not so well-known is that if the weight is uniformly distributed in the corresponding horizontal direction (that is, so that the weights of arcs of the cable are proportional to the lengths of their vertical projections on a horizontal plane) then the cable hangs in a parabola. This latter situation is very closely approximated by the cables of a suspension bridge supporting a very heavy horizontal roadbed by means of equally spaced suspenders, where in comparison with the roadbed, the weights of the cables and suspenders can be considered as negligible. It follows that the curve of a cable of a suspension bridge is (very closely) parabolic.

Now the shape of a conic is determined by its eccentricity. That is, all conics of the same eccentricity are similar, differing from one another merely in size. Since all parabolas have eccentricity 1, all parabolas are similar to one another, differing from one another only in size. Thus I was easily able to verify the parabolic nature of a cable of the George Washington suspension bridge by merely moving my parabolic template to and from my eye until it seemed to fit exactly along the curve of the distant cable (see Figure 1).

There, with the reader having patiently borne with me while I enjoyed giving my little lecture, I will continue with the story of our Hudson River ramble.

We walked several miles up river, failing to find Professor Kasner's hiding place. There we ate a frugal lunch that we had carried with us, and started our return trip to the George Washington bridge. After proceeding a distance, we decided to make

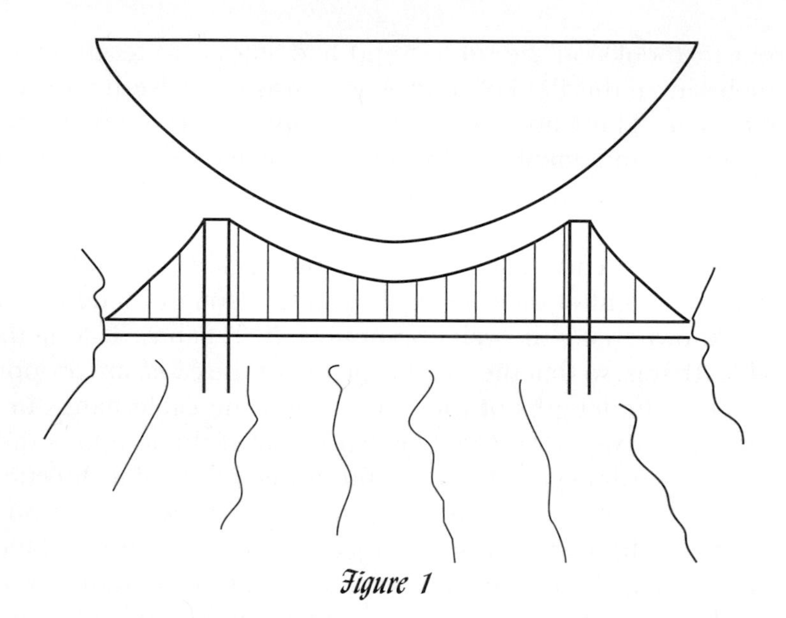

Figure 1

our way to the top of the Palisades and complete our return to the bridge along the edge of that lofty height. After we got to the top we soon realized that the plan was perhaps an ill-advised one, for we found ourselves trespassing across wealthy properties and climbing fences or working our way through hedges that separated the various properties from one another. We always disliked, and tried to avoid, trespassing, but being now committed we forged on, slinking along the outer cliff edge of the properties and trusting that we would not be discovered by anyone.

We seemed successful and at length came to a beautiful property with a lush green lawn running from a large house all the way to the cliff's edge. The day was warm and we settled down on the grass to rest a while and, we hoped, to escape the eye of a woman we had seen working in the flower garden in front of the house. While resting we noticed a remarkable thing. The lawn upon which we were sitting was literally covered with four-leaf clovers, scores and scores of them. As we were engaged in looking at this spectacle we were startled by a noise, and looking up saw the aforementioned woman bearing down on us, carrying a tray with some objects on it. We rose to our feet to meet the woman and to apologize profusely for trespassing. But she anticipated us and, smiling pleasantly, said, "I noticed you two young men resting here, and it occurred to me that you might like a glass of cold milk and some cookies. And she placed the tray containing these items on the grass beside us. We stammered some thanks, which she graciously waved aside.

I called the kind woman's attention to the remarkable abundance of four-leaf clovers in her lawn. "Oh yes," she said, "We've known about them for some time. They seem actually to increase in number each year. Why don't each of you pick several of them, take them home, and press them? Four-leaf clovers, you know, are said to bring good luck. And, by the way, don't worry about the tray and the glasses. Just leave them here on the lawn and I will gather them up later." We were almost speechless with gratitude. Would that all people of the world should be so kind and gracious to one another. I'm sure that neither Ted nor I will ever forget that wonderful woman and her understanding kindness. We decided each to pick a dozen four-leaf clovers and to take them home and press them as she had suggested.

Well, I have come to the end of my story about the mystery of the four-leaf clovers. Only two more details remain. On the bus traveling back to Paterson, Ted turned to me and in a very off-hand manner said, "By the way, Howard, I'm going to get married next week." Surprised I said, "You are? You sly old fox." I congratulated him and wished him the best. I went on and mused, "There is only one thing about it that bothers me. We shall never again go rambling together" "Oh yes we will," Ted emphatically replied. But my prophecy turned out to be the correct one. Not only did we never again ramble together, Ted moved to Newark and we have never seen one another again since that day. Which merely goes to prove, I suppose, that a congenial wife can easily supplant a mere rambling companion.

Finally, what has become of the four four-leaf clovers of the Dunnington text? I dearly hope that they are still there. In 1992 I gave my entire large mathematics library to the University of Central Florida. So, unless the four-leaf clovers have been disturbed, they should be still resting in the Dunnington volume, now reposing on one of the shelves of the University's library.

A Fugue

(*A talk delivered before a teachers' convention about 1963*)

(A *fugue* is a musical composition in which different parts are serially introduced, following one another in the manner of a flight or chase. The name derives from *fuga*, the Latin word for *flight*. The main parts of a strict fugue are an *exposition*, conducted by four voices, followed by a *discussion*. The exposition begins with the *subject* (theme) in the first voice. The second voice gives the *answer*, which is merely the subject transformed a fifth upward or downward. The third voice then introduces the original subject an octave below or above the principal key, and the fourth voice paralleling the second voice answers. The following discussion consists of less formal repetitions of the subject and answers.)

Exposition

First voice

There was man in Vermont who in his youth started an apple orchard from choice young trees. After years of carefully tending his developing trees, he found himself the possessor of a very impressive and lucrative orchard. His apple business so prospered that he found he needed a good-sized barn (instead of the inadequate shed he had started with), for storing his apples, keeping his ladders, baskets, barrels, and other picking gear, a small tractor for controlling the weeds and tilling the areas between his rows of trees, and a space to house a cider press and its allied equipment.

He began to inquire as to where he could find a suitable person to build him a barn. The country around had many old barns, almost all partially fallen down. He wanted a barn that, in his words, "would outlast the great pyramid of Cheops." When he stated his desire to neighbors and others, they all advised him to consult a man named Robert Grant, and they directed him to look at some barns that Grant had built. He did this and was greatly impressed by Grant's constructions; they were very sturdy and solidly built.

So the man looked up Grant and called on him one day to enlist his expertise in building him a barn. "Why is it," he asked Grant, "that your barns so outlast all other barns?" "It's an interesting story," replied Grant. "When I was in school I wanted to take the math course in trigonometry. I was so eager to take it that I skipped some of the preliminary math courses. To my disappointment I found I wasn't doing well in the course but I had a teacher who soon spotted my difficulty. She told me never to undertake anything without first establishing a sound foundation. It is foolish, she claimed, to build on a poor base. She directed me as to how to remedy my math foundation, and I became the school's top trigonometry student. I have always remembered that teacher's advice. The thing that makes my barns superior to so many other barns is simply the great care I devote to the foundations I put under them. So, you see, my success in barn building is because of a teacher I once had."

Second Voice

A man had an extensive lawn and found that he began to lack the time to take care of it himself. So he began seeking assistance. Now there were several lawn-carers who, with trucks carrying suitable equipment, like lawn mowers, edgers, weed-wackers, rakes, and other things, hired themselves out for such work. Inquiring from friends and others as to whom it would be good to get, he was most frequently advised to try to obtain Joe Lenner.

Now the man noticed a great deal of difference in the results of the various lawn-carers, and that Joe's work always showed a beauty and a finish generally lacking in the work of other carers. Joe would give careful attention to the best-looking mowing pattern for each lawn. His edging was complete and sharp, and he always took the trouble to protect the trunks of young trees before weed-wacking around them. He was finicky and his results were truly artistic compared to the slovenly and quickly-done work of many of the other carers.

So the man approached Joe to enlist him to care for his lawn. "Why is it," he asked Joe, "that you are so much neater and thorough in your work than so many other lawn-carers?" "That is a good question," replied Joe. "When I was in school I took a course in writing, and as my first assignment I was asked to write a brief biography of Abraham Lincoln. I rattled off a hasty paper, which I thought was adequate, and turned it in. When I got the paper back it had a grade of C– on it and a note from the teacher asking me to see her. I did so and she pointed out how I hadn't tried to do my best, that I had merely dashed off a quick paper—which was true. 'Now anyone can do that,' she said. 'Whenever you do a job you should try to do your very best. You should take pains and produce something of which you are proud.' I returned to the task and rewrote a biography of Lincoln that received high praise and was published in the school's annual magazine. So I might say my work is better than that of many other lawn-carers because of a teacher I once had."

Third Voice

I've always admired good carpentry. I particularly remember a fellow named Jim Hastings. When Jim undertook a carpentry job he would always bring with him some young boy, and all during the work he would take the time to explain carefully to the boy exactly what he was doing and why. In this way Joe passed on the basics and many of the fine points of carpentry to quite a number of young boys.

I got to know Jim well. One day I asked him, "Why do you take the time and trouble to instruct youngsters in your trade?

You could work much faster if you didn't do that." "Oh, yes," he replied, "I'm sure I could. But I remember a teacher who took extra time and pains with each of her students. 'You must never,' she maintained, 'hoard your knowledge or expertise just to yourself. You should always try to pass it on to someone else. In this way you can play a part in bettering the lives of others.' Her philosophy certainly bettered my life, and made a deep impression on me. So I suppose I find it good to share my carpentry skills because of a teacher I once had."

Fourth Voice

Mr. Brice had been a successful business man and when he retired he was able to purchase a large piece of property in his town. The property was in the form of a long isosceles triangle, the apex of which was bordered by two converging streets. He built his home at the base end of the triangle, and spent much time beautifully landscaping the property, especially the large tapered tip of it. Here he planted a number of shade trees, built a big rose arbor, and constructed a small pool with a gurgling fountain in it and colorful flower gardens surrounding it. He then placed several comfortable benches in shady nooks. It was a very attractive place and got to be known as Brice's Park.

Not long after Mr. Brice finished his little park, he attended a town meeting, and during the session he gave his park to the town, along with a monetary gift of such a size that the interest on it would cover the yearly upkeep of the park.

After the meeting was over, a friend of his asked him what prompted him to such a generous act. "I think the story will interest you," he said. "When I was in school I had a teacher who stressed that, should any of us become successful, we ought to share some of our good fortune to benefit others. She mentioned that is what any successful doctor should do with his medical knowledge, and that is what any successful art collector should do with his acquisitions, and so on. I thought the idea a fine and noble one, and I am merely trying to follow it in my own small way. I suppose, then, I might say I have given my little park to the town because of a teacher I once had."

Discussion

The King of a certain faraway country once decided to honor, before a public gathering, one of his subjects who had contributed the most to the good of the country. When the day of the affair arrived, a number of candidates appeared and presented their various credentials for the coveted honor.

There was the engineer who had designed and carried out the very useful and enviable transportation system of the country. There was the nurse whose untiring efforts had brought forth the country's unexcelled health system, with its great medical school and its fine hospital. There was the landscapist who had planned and created the country's beautiful and extensive park system. And there were a number of other equally outstanding candidates.

The King's forehead furrowed with perplexity over the difficult decision he had to make. As he was pondering on the matter he noticed a little gray-haired woman standing just behind the contestants. Turning to one of his courtiers he asked, "Who is that little gray-haired woman standing down there?" "Oh Sire," replied the courtier, "she is nobody." "But she seems so very interested in the proceedings," continued the King. "Sire," said the courtier, "she's naturally interested in the outcome of the contest." "But why?" asked the King. "Well you see, Sire, she was a teacher of all these contestants in their younger years, and did much to shape their future destinies. So she is naturally interested in which one will win the coveted honor." "Ah," said the King. His furrowed forehead cleared, and with no more hesitation he descended from the dais, walked past the gathered contestants, and proudly placed the wreath of honor on the gray head of the astonished old teacher.

Note that the first two voices above follow advice made by a teacher for benefiting the student, whereas the third and fourth voices follow advice made by a teacher for benefiting others than the student. In this way we carry out the octave difference between the first two and the last two voices.

Tombstone Inscriptions

A group of scholars, composed of a mathematician, a physicist, a chemist, a biologist, a novelist, and a school teacher, were assembled at a lunch table when the mathematician said, "Suppose, like Archimedes' request that the geometrical figure that led him to the discovery of the formulas for the area and volume of a sphere be engraved on his tombstone, what might each of us wish to have inscribed on our tombstones? I think I would like the figure of my simple 'proof without words' of a complicated trigonometric identity to be inscribed on my tombstone." The physicist said he would be pleased to have his famous aerodynamics equation inscribed on his. The chemist wanted a highly useful chemical formula that he had discovered on his. The biologist wished to have on his the name of a famous vaccine that he had created. The novelist wished to have on his tombstone the title of his book that won a Nobel Prize in Literature. Then, turning to the school teacher, the novelist, anticipating some fun at the teacher's expense, asked "And what would you like to have inscribed on your tombstone?" Without any hesitation the teacher replied, "I would be proud to have inscribed on my tombstone the names of my four students who most distinguished themselves in later life." Somehow the anticipated fun failed to materialize.

The Two Lights

It was around 300 BC that two lights were simultaneously lit in ancient Alexandria.

One of these lights was a physical light in the form of the world's first lighthouse, designed to guide ships coming down the Mediterranean Sea into the Great Harbor of Alexandria. Prior to this there had been large bonfires on shore to help guide navigators, but this was the first lighthouse in the modern sense of the word. It was built on the eastern end of a long island that lies closely off the coast of Egypt. The island was named Pharos Island, and so the lighthouse became known as the Pharos Light. Since the western end of the island was closer to the mainland, a great causeway was built there connecting the mainland to the island. It was across this causeway that the great blocks of stone were carried for the construction of the lighthouse. The structure was built by every artifice known to man at the time to outlast the ages; it was never to disappear from the face of the earth. When completed it rose to a height equal to half that of the Empire State Building in New York City, the height of the Washington Monument in Washington, D.C., the height of Diamond Head in Honolulu. Its great wood-burning light at the top could be seen some thirty-four miles out to sea. There has never since been built, in any part of the world, a lighthouse of such enormous height as that of the great Pharos Light.

At the same time that the great physical light was lit, an intellectual light was also lit in Alexandria. This light was in the form of thirteen rolls of papyrus, written by a patient and modest scholar named Euclid, and contained some 295 propositions in plane and solid geometry, elementary geometric algebra, and basic number theory. Whereas the great physical light shown for all who beheld it, the intellectual light shown for only those who sought it.

Now what has been the fate of those two lights? About 800 years ago the earth shook, and the great Pharos Light fell into the sea, and the subsequent ceaseless wash of the waves and shifting of the shoals have removed every trace of it. So of the great physical light, which was to outlast the ages, today no trace remains.

What about the concurrently lit intellectual light? It still shines brighter than ever, and not only in that part of the world where the Nile River issues upon the Mediterranean Sea, but in every part of the world wherever man seeks guidance from the human intellect. Of the two lights, the physical one and the intellectual one, the latter has far outlasted the former.

Now, throughout this world of ours, there is a craft known as teachers. What are these teachers daily doing in their classrooms? They are lighting little intellectual candles. How flimsy and how easily extinguishable these lights seem to be. But no, this is not so. For history has shown that of all the lights in the world, the intellectual lights are far and away the least extinguishable and the most lasting. So, dear teachers, should a day arrive, and it surely will, when you wonder if what you are doing in your classrooms is really worthwhile, when the legislature, and sometimes even parents, seem to fail to understand what you are doing, take heart, for some of the little candles that you are so patiently lighting will one day grow brighter and brighter and become brilliant lights that will benefit all mankind.

MMM

In a later reminiscence I will tell how an episode narrated by Florian Cajori in his *A History of Mathematics* led to a triple subscription to The Scholar's Creed. In the present reminiscence I shall tell how another episode narrated by Cajori in his book led to the creation of a remarkable collection entitled MMM. Let me explain.

In his book Cajori comments on Napoleon Bonaparte's interest in and flair for geometry. There is now a theorem and a construction problem each bearing Napoleon's name. Napoleon did not originate either the theorem or the construction problem; it was his popularization of them that caused his name to be attached to them. The theorem is a very pretty one and asserts that the centers of the three equilateral triangles described exteriorly on the sides of any given triangle constitute the vertices of a fourth equilateral triangle. The construction problem asks one to construct, with compasses alone, the vertices of a square inscribed in a circle of given center and radius.

Constructions employing only the compasses, with no use made of the straightedge, have become known as Mascheroni constructions, since the Italian geometer and poet Lorenzo Mascheroni (1750–1800) in 1797 published a detailed treatment of such constructions. Napoleon personally knew and admired Mascheroni, and it was from Mascheroni that he obtained the above problem. On p. 268 of Cajori's history book we read: "Napoleon proposed to the French mathematicians the problem, to divide the circumference of a circle into four

Florian Cajori, 1917

equal parts by the compasses only. Mascheroni does this by applying the radius three times to the circumference; he obtains the arcs *AB*, *BC*, *CD*; then *AD* is a diameter; the rest is obvious."

In reading the above, when I was in high school, it seemed to me that what Cajori had described was the obvious part of the desired construction, for any beginner in geometry would recognize that *A*, *B*, *C*, and *D* are four successive vertices of a regular hexagon inscribed in the given circle, making *A* and *D* a pair of diametrically opposite points. But how does one go about finding the midpoints of the two semicircular arcs formed by *A* and *D*?

This part of the construction Professor Cajori disposed of by merely saying it is "obvious." It wasn't "obvious" to me, and it took me a while to figure out a way of locating those midpoints using only the compasses. An interested reader might care, at this point, to try to supply Cajori's "obvious" conclusion.

In view of Professor Cajori's words, I felt I must be obtuse. So I finally sat down and wrote Cajori a letter (he was located in Colorado at the time), telling him how much I enjoyed his history book and that with a bit of effort I had found a way to do the "obvious" part of the construction problem described on p. 268, but that my construction certainly was not obvious. Would he kindly enlighten me?

I mailed off the letter and then waited and waited for a reply. Since a considerable time passed and I received no reply, I thought that perhaps university professors do not bother to answer letters from school children. But, finally, a reply to my letter did arrive. I can still (quite accurately, I think) recall the closing sentences of that reply. They ran: "If you know how to do the 'obvious' part of the construction problem on p. 268 of my book please tell it to me. I've neither eaten nor slept since I received your letter. Why I ever said the rest of the problem is 'obvious' I'll never know."

Well, you can imagine my surprise. Now I wasn't the promptest of correspondents in those days, and I let a while go by before I got around to writing out my construction of the 'obvious' part of the problem. It was in 1930, and I was still in high school. I laid the sealed and stamped (with a two-cent stamp) envelope on a table in the hall to be put out for the postman on the morrow and went into the living room to listen to the radio (this was in pre-television days). A news program was in progress, and to my astonishment I heard the newscaster announce: "Professor Florian Cajori, the well-known mathematician and historian of mathematics, died yesterday."

I had dallied too long preparing my answer to Professor Cajori's letter. So all these years now, I have had it on my conscience that I may have, albeit unintentionally, killed Professor Cajori. Clearly he had died of malnutrition and insomnia brought on by my tardiness, for he told me he had neither eaten nor slept since he had received my letter.

Professor Cajori's despairing letter lingered in my mind. After a while I decided to keep the letter and, as time went on, to add to it any other interesting mathematical items that might come my way. Thus was born a collection that I christened My Mathematical Museum, or, more briefly, MMM. I decided to number the items of the museum in the order of their acquisition. Professor Cajori's despairing letter became Item 1 of MMM. It would require a very bulky article to describe all the items finally acquired by MMM, for in time it contained over two thousand. We might, however, take the space to comment briefly on a few of the more interesting ones.

Postage Stamps

Some time ago I had a student from Norway enrolled in my Differential Equations course. Chatting with him one day I learned that his father worked for the Norwegian Postal Service. I asked the student if his father could get me specimens of the postage stamps that Norway had issued honoring that country's greatest mathematician, Niels Abel. The father graciously obliged, and I received samples of all the Abel stamps. With these stamps as a start I began to acquire other postage stamps bearing portraits of mathematicians upon them. Before the dissolution of MMM (described below), the museum contained postage stamps bearing portraits of fifty-three different mathematicians. Several mathematicians were represented by more than one country, and some, like Abel, were represented more than once by the same country. Russia and France were the most generous in representing mathematicians on postage stamps. England had never done so and the United States only once. Da Vinci, Galileo, Copernicus, and Einstein were each represented by four or more different countries.

The stamp issued by Norway in honor of Niels Henrik Abel. The stamp appeared in four colors: green, brown, red, and blue, each for a different amount of postage.

Portrait of Niels Henrik Abel

In addition to the above portrait stamps, MMM contained the five postal stamps The Netherlands issued in 1970 containing designs made with a computer coupled to a plotter, and also the set of stamps Nicaragua issued in 1971 paying homage to the world's "ten most important mathematical formulas."

Linkages

Upon describing to my older brother the Peaucellier Cell—the famous 7-bar linkage mechanism for drawing straight lines—he made me an elegant blackboard model of one. It was about a yard long when closed, had suction cups for the two fixed points, and a little hole for the insertion of a stick of chalk for describing the straight line. In 1874, exactly ten years after Peaucellier devised his mechanism, the Irish mathematician Harry Hart discovered a 5-bar linkage for drawing straight lines. So, inspired by my brother's model of the Peaucellier Cell, I constructed a blackboard model of my Hart's contraparallelogram. No one has been able to reduce the number of links in a linkage for drawing straight lines to fewer than five, or to prove that such a reduction is impossible. The so-called minimal linkage problem—that of determining the least number of links needed for drawing a given curve by a linkage mechanism—has never been solved for any curve other than the circle (1 link) and possibly the lemniscate of Bernoulli (3 links). It has been proved, however, that there exists a linkage for drawing any given algebraic curve, but that there cannot exist a linkage for drawing any transcendental curve. MMM finally contained blackboard models of linkages for drawing conics, cardioids, lemniscates, cissoids, and several other curves—a collection of thirteen linkages in all.

It is interesting that linkages have been devised for mechanically solving algebraic equations. Although MMM did not possess such a linkage mechanism, it did possess an entirely different type of mechanism for this purpose based upon reflections in mirrors.

Portraits

Over the years I collected for MMM ninety-two portraits of mathematicians, in some instances more than one portrait of a particular mathematician. Beyond a few extra large pictures, these portraits were all uniformly framed, under glass, in 12″ × 15″ thin black frames. Of these portraits seventy-five now appear in the 6th edition of my *An Introduction to the History of Mathematics*. The chief source was the magnificent David Smith Collection at Columbia University. These were augmented in a few instances by portraits from the New York Public Library, Culver Service, Brown Brothers, the Library of Congress, the Granger Collection, and the Bryn Mawr College Archives. MMM also contained the majority of these portraits reproduced, on a somewhat smaller scale, as transparencies for an overhead projector.

Polyhedra and spatial models

In time MMM possessed an enviable collection of scores of attractive polyhedral and spatial models, each at least a foot in diameter. Many of these models were constructed by my friend and former student Stephen Turner, and some others by various interested students of mine. The collection started, of course, with the five Platonic solids, and was then augmented by the Kepler-Poinsot solids, the Archimedean solids, the stellated polyhedra, and many special polyhedra and spatial models, like the Lennes polyhedra, examples of orientable and nonorientable manifolds, one-sided and two-sided manifolds, paradromic surfaces, map-coloring models used in combinatorial topology, flexagons, Jennifer's puzzle, braided polyhedra, pop-up polyhedra, rotating rings, collapsoids, zonohedra, and a few entertaining models like "passing a cube through a cube of the same size."

Personal items

No museum would be complete without some personal items associated with special people. MMM did not lack in this respect, for it contained, as examples more or less surreptitiously obtained, a lock of Einstein's hair, a long white woolen scarf worn by Professor G. H. Hardy, a sharp yellow pencil employed by Oswald Veblen, a walking stick once used by Paul Halmos, and others. And, of course, there were a number of letters and autographs of mathematicians. Many of these personal items have interesting stories connected with them. I always regretted that MMM did not possess a watercolor painting by Arthur Cayley, but all efforts to secure one failed. It would also have been nice to have had an original poem of Sir William Rowan Hamilton, penned in his own handwriting. There was, though, a spoof of Hamilton's historic scratching of the quaternionic multiplication table on one of the stones of the Brougham Bridge, humorously faked by one of my Summer Institute classes as having been blasted from the bridge and mailed, with genuine Irish stamps, to me from Ireland.

There are many other categories in MMM, like the extensive one devoted to mathematically motivated designs. We cannot, however, here go further into the contents of MMM but must hasten on to report what finally became of the collection.

Most of the items of MMM were acquired while I was a surveyor or while I was filling a series of transitory teaching posts. With no place to keep the growing collection I accepted the kind offer of Waldo, a friend of mine who occupied a very large old house on the outskirts of Pittsburgh, to store the stuff temporarily in his huge attic. So, every now and then as I flitted across the country in my station wagon, I would drop off accumulated loads at Waldo's house. There the great bulk of MMM rested until I finally settled permanently at the University of Maine, where I hoped in time to move the museum.

Such a move became urgent when I received a letter from Waldo telling me that he had managed, at a good price, to sell his old house. It was to be demolished to make way for a bypass around Pittsburgh. What should he do with MMM? I accordingly gave a talk to the University of Maine Mathematics Club, with the result that some of the students volunteered to serve as drivers of a couple of U-Drive-It trucks to bring the collection to Maine. There started an energetic but vain search for a place at the University to house the great collection. But conditions on the campus at that time were overcrowded and space simply nonexistent. Meanwhile Waldo's quandary was becoming critical.

So I finally sent Waldo an ad to run in the Pittsburgh papers, offering the museum's material to any teachers or supervisors of mathematics free for the carrying of it away. Waldo added his address and the times he would be available. It seems the little ad attracted wide attention, and Waldo reported that chunks of the museum were quickly and delightedly taken away. The postage stamps and the large collection of portraits went almost immediately. Most of the material was to find a place in schools and colleges.

I sometime feel sad and almost shed a tear when I think of the breaking up of MMM, for I know I could never rebuild it. But then I assure myself that it is certainly much better to have MMM scattered among a bunch of schools and colleges than to have it unused and collecting dust in some storage place.

Most of MMM moved westward, as testified by letters from teachers and professors who acquired parts of it. I recall an instance of this westward move, when some years later I gave a mathematical address at the College of Wooster. After the talk a schoolteacher, who had been in the audience, asked me if, while I was in Wooster, I would address his high school mathematics class. I agreed to do so, and imagine my surprise when I got to his school to find all four of the upper walls of his classroom occupied by a large number of my framed portraits of mathematicians. I asked him where he had obtained the portraits. He replied that they were already there when he inherited the classroom from his predecessor.

Of course, once you are a collector you will remain a collector, so I usually have a few recent interesting mathematical items on hand. But I manage to dispose of them rather quickly among the many visitors I have each summer at my retirement home in Maine. Indeed, I recently managed to clear out all lately accumulated curios when I gave my entire mathematics library to the University of Central Florida.

Completion of Napoleon's Problem

Let O be the center of the given circle. Draw the circles $A(C)$ and $D(B)$ to intersect in M. Now draw the circle $A(OM)$ to intersect the given circle in E and F. Then A, E, D, F are the vertices of an inscribed square.

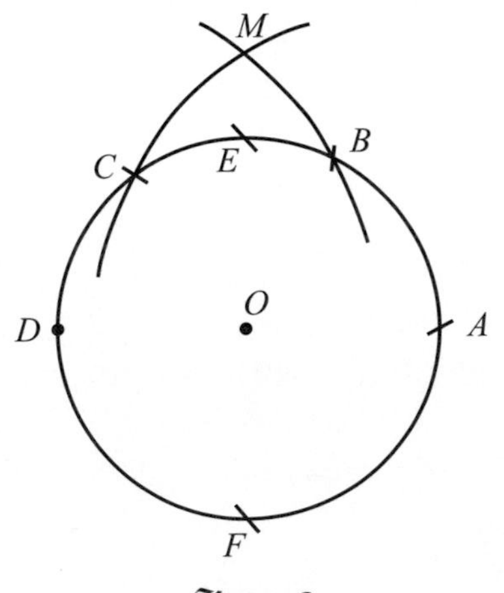

Figure 2

A considerably more challenging problem is the construction, with compasses alone, of the vertices of a square inscribed in a given circle where the center of the circle is not given.

Acquiring Some of the Personal Items for MMM

I mentioned, in the tale devoted to MMM, some of the personal items that ended up in that collection. It may be that certain readers are interested in how those items were secured. Here is the story of a few of them.

Professor Wiener's Hat

When I was a graduate student at Harvard University, a friend and I sought a ride to New Haven to attend a mathematics meeting held at Yale University. In those days very few students possessed cars, so we went about contacting professors and finally secured transportation in the car driven by Professor Norbert Wiener of MIT. Professor Wiener and Professor Dirk Struik (also of MIT) rode in the front and we two students rode in the rear. It was a very hair-raising ride, for the car wove perilously from one side of the road to the other as Professor Wiener, with his hands seemingly more often off the steering wheel than on it, energetically gesticulated during his conversation with Dr. Struik.

We finally arrived, miraculously safe, in New Haven. The day was rainy and we all wore hats to keep the rain from running down our necks. About noontime our group walked over to a cafeteria at Yale for lunch and we hung our raincoats and hats on a clothes tree in the front part of the cafeteria. After lunching, we rose to don our hats and coats. My friend and I politely stood back to let the two professors first secure their garments. Professor Wiener unthinkingly picked my hat (which in no way resembled his own) and placed it on his head. Since my hat was considerably smaller than his, I felt sure he would notice the error, but he didn't, and he stood there with my small hat ridiculously balanced on the top of his big head. I accordingly selected his large hat and put it on my head, it coming down over my ears and resting on the bridge of my nose. Thus capped I looked up at Professor Wiener, feeling certain that now he would see that something was wrong. He noticed nothing, not even the loud mirth of the others of our group, and thus attired we marched out of the cafeteria with traded hats.

There was a curious sequel to this escapade. After the mathematics meeting was over, my friend and I decided to take a safer ride back to Cambridge, and apparently Dr. Struik also returned by a different means. The next morning in Cambridge, listening to the news over the radio, I heard the newscaster report that the car of Professor Norbert Wiener of MIT was stolen from his garage while he was attending a mathematics meeting at Yale University. Professor Wiener had forgotten he had driven down to New Haven and had returned to Cambridge by bus. When he went to his garage the next morning, to secure his car to drive to work, he found it wasn't there and accordingly reported it as stolen.

A Lock of Dr. Einstein's Hair

I had the very good fortune to spend a year of graduate study at Princeton University. It was during that year that Dr. Albert Einstein became a member of the Institute for Advanced Study at Princeton, and I got to know the great man. Since I lived at the top of a short tower in the graduate school, and Dr. Einstein lived in the village on Mercer Street a couple of blocks beyond, I occasionally walked Dr. Einstein to or from the university. One day, with the rain lightly misting down, I called on Dr. Einstein at his home to accompany him to Fine Hall. The great scientist managed to escape from the house without his hat, on the assurance to his wife that "My hair will dry faster than my hat will." During the walk to the university, the rain managed to batten down some of this considerable mane over his forehead. We proceeded to his office in Fine Hall, discussing something, and he let his hair dry in its annoying position. Bothered by the hair, he would every now and then swipe at it with his hand in an effort to brush it up, but it only fell back over his forehead. Suddenly, in a little fit of petulance, he stalked over to a small desk, pulled out the drawer, secured a pair of scissors, and cut off the offending lock of hair, throwing it in the waste basket. Always thinking of my mathematical museum, I later returned to the office, extracted the lock of hair from the basket, and reverently put it in an envelope.

Professor Hardy's Scarf

During my year of graduate study at Princeton University, Professor G. H. Hardy came over from England to spend a semester lecturing on some Tauberian theorems. He soon had us all charmed.

Hardy wore heavy tweed suits, and even in the coldest weather augmented his dress with only a very long dirty white scarf that he wrapped round and round his neck. It wasn't long before I coveted that scarf for my mathematical museum. So, one day, after Professor Hardy had gone into the great Gothic dining hall where we ate suppers, I took his long scarf off the hook where he had hung it and replaced it with the nicest and most expensive scarf I had been able to purchase in the village. After supper, when Professor Hardy came out to reclaim his scarf, he found the beautiful one in its place. He began grumbling, but I soothed him by pointing out that whoever had taken his scarf had certainly left a considerably nicer one in its place. He conceded this, and finally took the new scarf in place of his old one. I waited several days to be sure he was satisfied with the surreptitious exchange. Finding him quite content, I entered his scarf as a new acquisition into my mathematical museum—in its original dirty and unkempt condition, of course.

A Doubtful Acquisition

It was the summer of 1960 that the great American geometer Oswald Veblen died at his summer cottage near Brooklin, Maine. The following summer I visited the cottage, much as one would visit a sacred shrine, and there I met Mrs. Veblen, who was about to take some friends on a little automobile excursion. She graciously gave me permission to wander about the grounds while she was gone, and even left the cottage unlocked in case I wished to explore inside. I declined the invitation to go inside, but I did walk around the rather extensive grounds, from the cottage, which was set well in from the road, down to the seashore, where the view of distant Acadia National Park was superb. In my wandering about, I came across a full-length bright yellow lead pencil, broken in the middle but still adhering together, as though it had been stepped upon and lain on the ground all winter. I picked the pencil up, musing to myself as to whether it might have belonged to and been used by the master himself. For, I reasoned, a geometer almost always carries a pencil with him, and if Mrs. Veblen were to use a pencil she would in all likelihood use it indoors to write out a grocery order or some such thing. At length I pocketed the pencil as a possible acquisition for my mathematical museum.

The next summer I happened to teach in an NSF Summer Institute at Bowdoin College, and Professor Albert Tucker of Princeton was also there. During a luncheon together I told Professor Tucker the story of the broken pencil. He immediately settled into a concentrated study, apparently trying to figure out how he might discredit the acquisition. Finally, brightening up, he asked me a startling question. "Did you know," he triumphantly asked, "that during his last years Professor Veblen was almost totally blind—that he possessed only a peripheral vision?" I hadn't known this, and for a moment it looked as

Oswald Veblen

though the pencil would have to be discarded. But then it occurred to me: who but a person with only peripheral vision would drop a full-length bright yellow pencil on the ground and then not be able to find it again? When I disclosed this line of reasoning to Professor Tucker, he subsided into a seemingly disappointed silence.

Difficulty in Explaining Relativity Theory in a Few Words

Reminiscing is often like a chain reaction. One reminiscence will recall another, and then that other will recall a still further one, and so on.

The mention of Mrs. Veblen in the previous reminiscence recalled to my mind the weekly afternoon teas she used to give, at her home in Princeton, for the local mathematicians and their friends. These were very pleasant affairs, and gave the mathematicians an opportunity to talk a little "shop" among themselves.

Mrs. Veblen dearly wished that at some time Dr. Einstein would attend one of her teas, but Dr. Einstein was not keen about such social gatherings. Imagine, then, Mrs. Veblen's surprise and joy when, at one of her teas, Dr. Einstein suddenly appeared and sat down. She was so excited that she could hardly contain herself. She crossed the room fluttering her arms for silence, and then remarked: "We are particularly honored this afternoon to have Dr. Einstein with us." Then, turning to Dr. Einstein she said: "I wonder, Dr. Einstein, if you would be so kind as to explain to my guests in a few words, just what is relativity theory?"

I wondered what the poor man was going to do, but without any hesitation Dr. Einstein rose to his feet and commenced a story. He said he was reminded of a walk he one day had with his blind friend. The day was hot and he turned to the blind friend and said, "I wish I had a glass of milk."

"Glass," replied the blind friend, "I know what that is. But what do you mean by milk?"

"Why, milk is a white fluid," explained Einstein.

"Fluid, now I know what that is," said the blind man, "but what is white?"

"Oh, white is the color of a swan's feathers."

"Now, feathers, I know what they are, but what is a swan?"

"A swan is a bird with a crooked neck."

"Neck, I know what that is, but what do you mean by crooked?"

At this point Einstein said he lost his patience. He seized his blind friend's arm and pulled it straight. "There, now your arm is straight," he said. Then he bent the blind friend's arm at the elbow. "Now it is crooked."

"Ah," said the blind friend. "Now I know what milk is."

And Einstein, at the tea, sat down.

Difficulty in Obtaining a Cup of Hot Tea

Continuing the chain reaction set up in the previous two reminiscences leads me to the following little story. Again it concerns G. H. Hardy.

At the top of Fine Hall (Princeton's mathematics building) was the university's great mathematics library. Around the outer edges of the library were a number of dormer nooks, each equipped with a chair and a desk. These nooks were for the use of the graduate mathematics students and visiting mathematics scholars. I was assigned one of these nooks and Professor Hardy was given one.

Now every afternoon, at four o'clock, there would be a soft tinkle of a bell. This signified that there were refreshments—ice cream, coffee, tea, milk, sandwiches, etc.—downstairs in the large Common Room. Most of us in the library were interested in these refreshments. But when Professor Hardy was there, we got to know that as soon as the bell tinkled, we had better lean well over our desks to hold down our papers, for like a flash Professor Hardy would streak by and create a vacuum behind him that would suck everything right off your desk. By the time we would get downstairs to the Common Room, we would find Professor Hardy standing very disconsolately in a corner of the room. So one day I thought, "I'm going down early and see what that man does, and why he's so eager to get down there."

So I kept careful track of time, and I went down several minutes before the little bell was to tinkle upstairs, and I saw the

G. H. Hardy in his rooms at Trinity College, Cambridge

lady put her finger on the bell button. She had hardly pulled her finger away when in raced Professor Hardy. He tore over to the big silver teapot, put the back of his hand on it and said, "Aah, stone cold," and then took up his disconsolate position in a corner of the room.

I went over and put the back of my hand on the teapot, and then walked over to the infirmary to be treated for a burned hand. Professor Hardy must have had asbestos hands to put them on that teapot. He claimed that he couldn't get a cup of hot tea in America.

Hail to Thee, Blithe Spirit!

Sketch by Dave Logothetti.

The great dream of every teacher of mathematics is to find among his or her students a potential future mathematician, and to play a beneficial part in nourishing the mathematical growth of that student. There cannot be many joys that exceed such an experience. I have been very lucky in this aspect of teaching and I could easily draw up quite a list of former students who have given me this supreme joy. I doubt they realize how deeply I thank them for having been students of mine. I have selected the following reminiscence to illustrate this aspect of teaching, an aspect that makes many of us profoundly happy to be in the profession, and that makes all the frustrations and problems of the profession seem unimportant.

The sudden death of the mathematician Verner Hoggatt, Jr., who, among many other things, was the founder of *The Fibonacci Quarterly*, led, in August 1951, to the publication of a Memorial Issue dedicated to Vern. Because of my long association with Vern, first as his teacher and then as a collaborator and friend, I was invited to write the opening article for the issue. Herewith is that article.

It was in the mid-1940s that I left the Department of Applied Mathematics at Syracuse University in New York State to chair a small Department of Mathematics at the College of Puget Sound[1] in Tacoma, Washington. Among my first teaching assignments at the new location was a beginning class in college algebra and trigonometry. At the first meeting of this class I noticed, among the twenty-some assembled students, a bright-looking and somewhat roundish fellow who paid rapt attention to the introductory lecture.

As time passed I learned that the young fellow was named Verner Hoggatt, fresh from a hitch in the army and possessed of an unusual aptitude and appetite for mathematics. Right from the start there was little doubt in my mind that in Vern I had found the mathematics instructor's dream — a potential future mathematician. He so enjoyed discussing things mathematical that we soon came to devote late afternoons, and occasional evenings, to rambling around parts of Tacoma, whilst talking on mathematical matters. On these rambles I brought up things that I thought would particularly capture Vern's imagination and that were reasonably within his purview of mathematics at the time.

Since Vern seemed to possess a particular predilection and intuitive feeling for numbers and their beautiful properties, I started with the subject of Pythagorean triples, a topic that he found fascinating. I recall an evening, shortly after this initial discussion, when I thought I would test Vern's ability to apply newly acquired knowledge. I had been reading through Vol. I of Jakob Bernoulli's *Opera* of 1744, and had come upon the alluring little problem:

'Titius gave his friend, Sempronius, a triangular field of which the sides, in perticas, were 50, 50, and 80, in exchange for a field of which the sides were 50, 50, and 60. I call this a fair exchange." I proposed to Vern that, in view of the origin of this problem, we call two noncongruent isosceles triangles a *pair of Bernoullian triangles* if the two triangles have integral sides, common legs, and common areas. I invited Vern to determine how we might obtain pairs of Bernoullian triangles. He immediately saw how such a pair can be obtained from any given Pythagorean triangle, by first putting together two copies of

[1] Now the University of Puget Sound.

the Pythagorean triangle with their shorter legs coinciding and then with their longer legs coinciding. He pointed out that from his construction, the bases of such a Bernoullian pair are even, whence the common area is an integer, so that these Bernoullian triangles are also Heronian (that is, they have integral areas and integral sides).

On another ramble I mentioned the problem of cutting off in a corner of a room the largest possible area by a two-part folding screen. I had scarcely finished stating the problem when Vern came to a halt, his right arm at the same time coming up to a horizontal position, with an extended forefinger. "There's the answer," he said. I followed his pointing finger, and there, at the end of the block along which we were walking, was an octagonal stop sign.

There was a popular game at the time that was, for amusement, engaging many mathematicians across the country. It had originated in a problem in *The American Mathematica1 Monthly*. The game was to express each of the numbers from 1 through 100 in terms of precisely four 9s, along with accepted mathematical symbols of operation. For example

$$
\begin{aligned}
1 &= 9/9 + 9 - 9 = 99/99 = (9/9)^{9/9}, \\
2 &= 9/9 + 9/9 = .\dot{9} + .\dot{9} + 9 - 9, \\
3 &= \sqrt{\sqrt{9}\sqrt{9}} + 9 - 9 = \left(\sqrt{9}\sqrt{9}\sqrt{9}\right)/9.
\end{aligned}
$$

The next day Vern showed me his successful list. In this list were the expressions

$$
\begin{aligned}
67 &= \left(.\dot{9} + .\dot{9}\right)^{\sqrt{9}!} + \sqrt{9} = \left(\sqrt{9} + .\dot{9}\right)^{\sqrt{9}} + \sqrt{9}, \\
68 &= \sqrt{9}¡\left(\sqrt{9}!\sqrt{9}! - \sqrt{9}¡\right), \\
70 &= \left(9 - .\dot{9}\right)9 - \sqrt{9}¡ = \left(.\dot{9} + .\dot{9}\right)^{\sqrt{9}!} + \sqrt{9}!,
\end{aligned}
$$

where the inverted exclamation point, ¡ indicates subfactorial.[2] For all the other numbers from 1 through 100, Vern had been

---

[2] $n¡ = n!\left[1 - 1/1! + 1/2! - 1/3! + \cdots + (-1)^n n!\right]$.

able to avoid both exponents and subfactorials, and so he now tried to do the same with 67, 68, and 70, this time coming up with

$$67 = \sqrt{9!/(9 \times 9) + 9},$$

$$68 = \left(\sqrt{9}\,!\right)!/9 - \sqrt{9}\,! - \sqrt{9}\,!,$$

$$70 = \left(9 + .\dot{9}\right)\left(\sqrt{9}\,! + .\dot{9}\right).$$

It would take too much space to pursue further the many things we discussed in our Tacoma rambles, but, before passing on to later events, I should point out Vern's delightful wit and sense of humor. I'll give only one example. The time arrived in class when I was to introduce the concept of mathematical induction. Among some preliminary examples, I gave the following. "Suppose there is a shelf of 100 books and we are told that if one of the books is red then the book just to its right is also red. We are allowed to peek through a vertical slit, and discover that the sixth book from the left is red. What can we conclude?' Vern's hand shot up, and upon acknowledging him, he asked, "Are they all good books?" Not realizing the trap I was walking into, I agreed that we could regard all the books as good ones. "Then," replied Vern, "*all* the books are red." "Why?" I asked, somewhat startled. "Because all *good* books are read," he replied, with a twinkle in his eye.

It turned out that I stayed only the one academic year at the College of Puget Sound, for I received an attractive offer from Professor W. E. Milne of Oregon State College[3] to join his mathematics staff there. The hardest thing about the move was my leave-taking of Vern. We had a last ramble, and I left for Oregon.

I hadn't been at Oregon State very long when, to my great joy and pleasure, at the start of a school year I found Vern sitting in a couple of my classes. He had decided to follow me to Oregon. We soon inaugurated what became known as our "oscillatory rambles." Frequently, after our suppers, one of us would call at the home of the other (we lived across the town of Corvallis

[3] Now Oregon State University.

from one another), and we would set out for the home of the caller. Of course, by the time we reached that home, we were in the middle of an interesting mathematical discussion, and so returned to the other's home, only to find that a new topic had taken over which needed further time to conclude. In this way, until the close of a discussion happened to coincide with the reaching of one of our houses, or simply because of the lateness of the hour, we spent the evening in oscillation.

Our discussions now were more advanced than during our Tacoma rambles. I recall that one of our earliest discussions concerned what we called *well-defined* Euclidean constructions. Suppose one considers a point of intersection of two loci as *ill-defined* if the two loci intersect at the point in an angle less than some given small angle q, that a straight line is *ill-defined* if the distance between the two points that determine it is less than some given small distance d, and that a circle is *ill-defined* if its radius is less than d; otherwise, the construction will be said to *be well-defined*. We proved that *any Euclidean construction can be accomplished by a well-defined one*. This later constituted our first jointly published paper (in *The Mathematics Teacher*). Another paper (published in *The American Mathematical Monthly*) that arose in our rambles, and an expansion of which became Vern's master's thesis, concerned the derivation of hyperbolic trigonometry from the Poincaré model. We researched on many topics, such as Schick's theorem, nonrigid polyhedra, new matrix products, vector operations as matrices, a quantitative aspect of linear independence of vectors, trihedral curves, Rouquet curves, and a large number of other topics in the field of differential geometry.

We did not forego our former interest in recreational mathematics. The number game of the Tacoma days had now evolved into what seemed a much more difficult one, namely, to express the numbers from 1 through 100 by arithmetic expressions that involved each of the ten digits 0, 1, ..., 9 once and only once. This game was completely and brilliantly solved when Vern discovered that, for any nonnegative integer n,

$$\log_{(0+1+2+3+4)/5}\left\{\log_{\sqrt{\sqrt{\cdots\sqrt{(-6+7+8)}}}} 9\right\} = n,$$

where there are n square roots in the second logarithmic base. Notice that the ten digits appear in their natural order, and that, by prefixing a minus sign if desired, Vern had shown that *any integer,* positive, zero, or negative, can be represented in the required fashion.[4]

A little event that proved very important in Vern's life took place during our Oregon association. When I was first invited to address the undergraduate mathematics club at Oregon State, I chanced to choose for my topic, "From rabbits to sunflowers," a talk on the famous Fibonacci sequence of numbers. Vern, of course, attended the address, and it reawakened in him his first great mathematical interest, the love of numbers and their endless fascinating properties. For weeks after the talk, Vern played assiduously with the beguiling Fibonacci numbers. The pursuit of these and associated numbers became, in time, Vern's major mathematical activity, and led to his eventual founding of *The Fibonacci Quarterly,* devoted chiefly to the study of such numbers. During his subsequent long and outstanding tenure at San Jose State University, Vern directed an enormous number of master's theses in this area, and put out an amazing number of attractive papers in the field, solo or jointly with one or another of his students. He became the authority on Fibonacci and related numbers.

After several years at Oregon State College, I returned east, but Vern continued to inundate me with copies of his beautiful findings. When I wrote my *Mathematical Circles Squared* (Prindle, Weber & Schmidt, 1972), 1 dedicated the volume

To VERNER E. HOGGATT, JR.
Who, over the years, has sent me more
Mathematical goodies than anyone else

The great geographical distance between us prevented us from seeing one another very often. I did, on my way to lectur-

---

[4] Another entertaining number game that we played was that of expressing as many of the successive positive integers as possible in terms of not more than three π's, along with accepted symbols of operation.

ing in Hawaii, stop off to see Vern, and I spent a few days with him a couple of years later when I lectured along the California coast. He once visited me at the University of Maine, when, representing his university, he came as a delegate to a national meeting of Phi Kappa Phi (an academic honorary that was founded at the University of Maine). For almost four decades I had the enormous pleasure of Vern's friendship, and bore the flattering title, generously bestowed upon me by him, of his "mathematical mentor."

In mathematics, Vern was a skylark, and I regret, far more than I can possibly express, the sad fact that we now no longer will hear further songs by him. But, oh, on the other hand, how privileged I have been; I heard the skylark when he first started to sing.

Hail to thee, blithe Spirit!
Bird thou never wert,
That from Heaven, or near it,
Pourest thy full heart
In profuse strains of unpremeditated art.

C. D.

No, the C. D. does not stand for "certificate of deposit," but for a dear friend of mine. Working, as I have, for such a long period of time with other mathematicians, I have acquired many mathematical friends. I could devote quite some space to telling about them. Prominent in any such account would be Clayton Dodge, now a University of Maine Emeritus Professor of Mathematics.

I first met Clayton shortly after I joined the mathematics staff at the University of Maine back in 1954. At that time Clayton was a sharp, keenly interested, and outstanding graduate mathematics student. I had the great pleasure of having him in several of my graduate classes, and in time he wrote his Master's Thesis with me and served as a very able assistant in some of my courses. When the University conducted some National Science Foundation Summer Institutes for Teachers of Mathematics, Clayton did yeoman work assisting the enrollees at the afternoon problem sessions. When I revised my book on *The Foundations and Fundamental Concepts of Mathematics,* I greatly benefited from Clayton's excellent suggestions and wise counsel. Meanwhile he was working his way up the teaching ladder at the University, constantly winning the deep admiration of his students. When my work as a Problem Editor of *The American Mathematical Monthly* grew beyond the bounds of a single person's time, I was able to secure Clayton's assistance. Clayton has become one of the top problemists of the country, and when Leon Bankoff retired as Problem Editor of *The Pi Mu Epsilon*

Photo by Dorothy R. Dodge

Clayton Dodge

Journal, it was natural that Clayton should be invited to succeed him. For some years now Clayton has conducted an outstanding problem section for that journal. Over the years Clayton has written several excellent mathematics texts, some of them particularly tailored to fit new courses that he introduced. His care of his students was very evident, and often, when travelling to a mathematics meeting, such as a session of the Northeastern Section of the MAA, he would take along a student or two.

Clayton is much more than just a mathematician. His fine voice can be heard in the offerings of The University of Maine Oratorio Society. He is also a skilled carpenter, decorator, and artisan, and is always engaged in an interesting project of one kind or another. A few years ago he made me a very remarkable large lamp in the form of a great stellated regular icosahedron. To make the lamp he had to cut sixty congruent equilateral triangular pieces of glass. The lamp is cleverly designed to allow the escape of the heat generated by the light bulb, and it has an ingenious hinged top permitting easy change of the bulb. This lamp has become the showpiece of my summer home in Lubec, Maine.

Coupled with all the above is Clayton's unique humor and wit, a few examples of which are appended at the end of this article. But it is the deep friendship that has developed between us that I treasure the most. I regard him as beyond any question one of my foremost friends. If a kind fairy should come along and offer to grant me any wish I might care to make, I would wish that everyone could have a friend like Clayton Dodge. My years of association with him have been one of the truly finer events of my life.

Myself, Clayton Dodge, and Albert Wooten—
the leaders of the 8 NSF Teachers Institutes at the University of Maine

Samples of Clayton's mathematical wit and humor

Continued fractions. Helmut Hasse spoke yesterday on continued fractions. But, of course, he didn't finish.

Clayton Dodge's rectangle (*wrecked angle*).

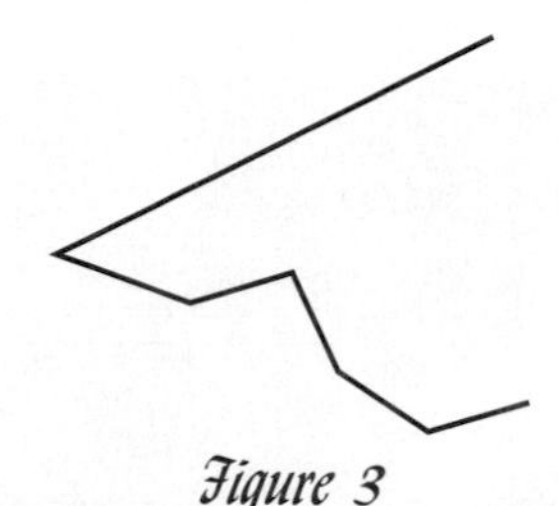

Figure 3

Cute. Question: Define *acute right triangle*. Answer: A right triangle with pretty legs.

A Fourier Series. Yea + Yea + Yea + Yea.

A footnote. On page 22 of Clayton Dodge's *The Circular Functions*, where circular coordinates and their notation are introduced, appears the footnote "Since it is customary to use *round* parentheses to enclose *rectangular* coordinates, it seems equally logical to use *square* brackets to enclose *circular* coordinates."

Pie in the bathroom. The digits in the decimal expansion of pi seem to satisfy no pattern and appear to occur in an absolutely random fashion. Because of this, the successive digits in the decimal expansion of π have been used where a random sequence of digits is required. Use of this randomness can be made in laying tile, where, to break the monotony of a uniform color, one wishes every now and then to insert a tile of a different color. Clayton Dodge employed this idea when tiling a wall of his bathroom in basic white tiles interspersed randomly with tan ones.

Cupid's Problem

For 25 years I had the enormous pleasure of serving as Editor of the Elementary Problems Department of *The American Mathematical Monthly,* succeeding Professor H. S. M. Coxeter of the University of Toronto when he retired from that position. One of the enjoyable features of the position lay in the resulting interesting correspondence with a large number of both domestic and foreign mathematicians. I got to know a great many of these mathematicians and the work that currently engaged them. But in time, with the huge growth of the membership of the MAA, it became essentially impossible for a single individual to handle the resulting burgeoning Problem Department correspondence. For a time I held on, with the valuable assistance of Professor Dodge. Finally, however, I found it wise to recommend that the department be handled by a group of mathematicians, the group containing a geometer, an algebraist, an expert on number theory, and so on, so that the proposals and solutions could be read and judged by an appropriate member of the group. The first group to undertake the task was located at the University of Maine, and it then passed on to a succession of similar groups located at other universities.

It was customary, of course, to acknowledge, with their locations, both the proposers and the solvers of the various problems, and to give a title to each problem when its solution appeared. Herein elements of mysticism occasionally arose. For instance, the solvers of a number of problems read: CHED, Stratford on Penobscot. The HE stood for Howard Eves and the CD for Clayton Dodge, and the University of Maine lies along the Penobscot River. An example of mysticism in the selection of a title to a problem is Problem E 740, which was entitled "Cupid's Problem." I imagine few readers of the journal were able to guess the reason why. Here is the explanation of this curious selection of a title.

Problem E 740 appeared in the October 1946 issue of *The American Mathematical Monthly,* and was: "Let there be given five points in the plane. Prove that we can select four of them which determine a convex quadrilateral." The problem was published as proposed by Esther Szekeres, then in Shanghai, China, and when solutions to the problem were published in the May 1947 issue of the journal, the problem was titled "Cupid's Problem." There is no doubt that very few readers of the *Monthly* understood why the problem received this peculiar title. The reason is this. The problem had been submitted, not by Esther Szekeres, but by a close friend of both Esther and her husband George Szekeres. The friend, who wished to remain anonymous, stated that it was this problem and some of its generalizations that first brought Esther and George together, and he asked that the problem be published as proposed by Esther. Cognizant of the above romantic facts, I was led to title the problem "Cupid's Problem."

The general case of the problem is discussed by Paul Erdős and George Szekeres in an article in *Compositio Mathematica,* vol. 2 (1935), pp. 463–470. It was known that from nine points in the plane we can select five which determine a convex pentagon, but it was not known if from $2^{n-2} + 1$ points in the plane, where n is an integer greater than 3, we can always select n which determine a convex n-gon. Given $n + 3$ points in an n-flat, we can select $n + 2$ which determine a convex polytope.

The Lighter Life of an Editor

Many amusing things occur over a long tenure of editorship, at least so it happened to me during my more than twenty-five years as editor of the Elementary Problems Department of *The American Mathematical Monthly*. Here are three instances of this lighter side of editorial work.

The Case of Mr. S. T. Thompson

The Thompson case came about in this way. On occasion I found in my editorial work that I was able to discover a considerably better solution to a problem than any of those that were submitted. Indeed, sometimes my solution was the only one I possessed. Feeling it uncomfortably forward to publish my own solution, I decided to invent a mythical character for the purpose, and, avoiding such obvious names as Mr. Ed I. Tor, I created a Mr. S. T. Thompson, placing him vaguely in Tacoma, Washington (where I was living at the time), and gave him credit for my solutions.

Now Professor Harry Gehman, who for many years was the loved, energetic, and superb Secretary-Treasurer of the Mathematical Association of America, was ever alert for possible new members of the Association. Seeing Mr. S. T. Thompson's name appearing now and then among the solvers of the *Monthly*'s problems, Harry wrote to me asking for Thompson's

Harry Gehman

complete address, as he wished to invite Thompson to become a member of the Association. I wrote back to Harry that there was no use in inviting Mr. Thompson to become a member, for Mr. Thompson had an insuperable dislike of the Association and absolutely nothing would ever induce him to join. Needless to say, Harry found this attitude puzzling, frustrating, and quite incomprehensible, and at a number of subsequent meetings of the Association he would approach me and wonder on Mr. Thompson's dislike of the Association.

It was not until 1971, when my book *Mathematical Circles Revisited* appeared, that I finally divulged the true state of affairs about Mr. S. T. Thompson.

Professor Euclide Paracelso Bombasto Umbugio

Another case, somewhat similar to that of Mr. S. T. Thompson, originated in the pages of the Elementary Problems Section of *The American Mathematical Monthly* back in 1946. At that time, Professor George Pólya and I thought it might be enlivening if in each April issue of the journal there appeared a sort of April fools problem—a problem for which a straightforward solution is tedious and long, but for which with cleverness one can find an extremely brief and elegant solution. It was decided that these problems would emanate from a windy, verbose, but kindly numerologist, Professor Euclide Paracelso Bombasto Umbugio of Guayazuela.

Professor Umbugio's fame spread and in time April fool proposals were submitted through him by good bona fide mathematicians. But the hoax fooled many readers of the *Monthly*, and the letters received asking for his address (so that some scientific correspondence could ensue) would fill a little pamphlet. Professor Leo Moser, in his graduate-school days, was among those seeking Professor Umbugio's address. All inquirers were informed that Professor Umbugio moved about so much that the best way to reach him was through correspondence sent in care of the editor of the Elementary Problems section.

George Pólya in the late 1930s

The Episode of the Butterfly Problem

There is an elusive and tantalizing problem in elementary geometry known as the "butterfly problem": *Let O be the midpoint of a given chord of a circle, let two other chords TU and VW be drawn through O, and let TW and VU cut the given chord in E and F respectively; prove that O is the midpoint of FE.* The problem receives its name from the fancied resemblance of the figure (see Figure 4) to a butterfly with outstretched wings. It is a very unusual student of high school geometry who succeeds in solving the problem.

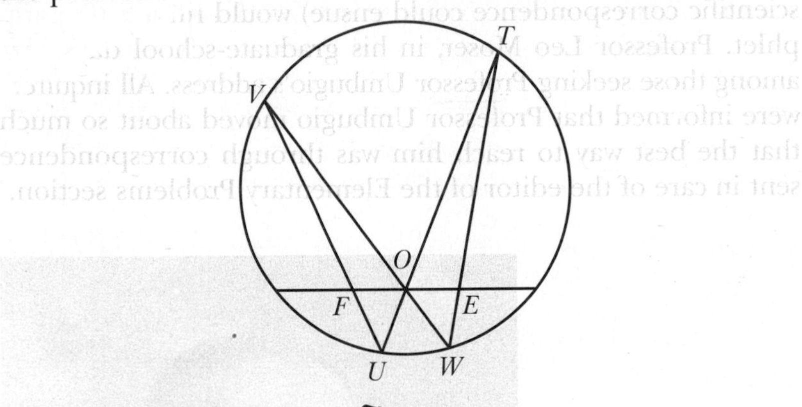

Figure 4

In an effort to obtain a number of different but elegant solutions to the butterfly problem, I once ran it in the Elementary Problems section of *The American Mathematical Monthly* (Problem E 571, May 1943). Some truly ingenious solutions were submitted, and in due time (the February 1944 issue) I published a number of them. Among the published solutions was a particularly attractive one by Professor E. P. Starke, one of our country's outstanding problemists and the editor, at the time, of the Advanced Problems section of the *Monthly*.

Then a few years passed by, and one day I received a letter from Professor Starke stating that that semester he had a bright Norwegian student in one of his classes who gave him a very troublesome geometry problem. Professor Starke confessed that he had spent an inordinate amount of time on the problem, but without success, and he wondered if I could perhaps help him out with it. He then stated the problem, and lo, it was the butterfly problem! For a solution I referred him, without mentioning any names, to those pages in *The American Mathematical Monthly* wherein his own elegant treatment appeared.

Note 1. The above episode of the butterfly problem recalls a similar occurrence in the life of J. J. Sylvester. Arthur Cayley and Sylvester, mathematical friends over a long period of years, were in almost every sense antitheses of one another. One of the many ways in which they were opposites was in the matter of memory. Cayley seemed never to forget anything he had once seen or read, whereas Sylvester had difficulty in remembering even his own mathematical inventions. On one occasion Sylvester objected to a mathematical statement made by a companion, insisting that the statement had never been heard of and, moreover, simply could not be true. The companion responded by showing the amazed Sylvester a paper written by Sylvester himself, in which Sylvester had announced his discovery of the concerned statement and had written out its proof.

Note 2. The butterfly problem is a fine example of those problems which the more they are generalized the easier they seem to prove. Thus, if in the butterfly problem we should replace (see the above figure) the given circle by an arbitrary given conic and the pair of lines TW and VU by another arbitrary conic through the four points T, U, V, W, a proof more readily falls out.

The Two Kellys

There are two Kellys, Leroy Kelly and Paul Kelly, each of whom has graced the American field of mathematics and, in particular, the area of geometry. Naturally, as geometers, I had corresponded with each Kelly for a number of years, but for a long time had not had the pleasure of meeting either one of them. The story of how I finally met these two mathematicians is interesting.

One summer in the 1960s I was teaching at the University of Maine in a National Science Foundation Summer Institute for Teachers of Mathematics. Driving home from class one day, as I came to my house in nearby Stillwater, I noticed a car parked in front of the house, with a man sitting in it. I drove up my driveway to the big barn in the rear, the lower floor of which served as my garage. Glancing in the car's rear-view mirror I saw the man of the parked car get out of his car and start walking up the driveway. So, after I parked my car, I proceeded down the driveway toward the advancing man. As we drew near he extended a hand, which I grasped, and he said, "I'm Kelly." "Ah," I replied, "the great geometer." "Oh, no," he said, "I'm not Paul Kelly, I'm Leroy Kelly." We went up into the top of my barn where I had a large study room that housed my mathematics library. There Leroy and I had a wonderful hour or more talking geometry. Leroy was, if I recall correctly, driving to the Canadian Maritime Provinces, and thoughtfully took in a visit with me on his way.

Only a year or so after meeting Leroy Kelly, I flew to Corpus Christi to attend a large mathematics meeting to which I had been invited to present a paper. Having earlier received a copy of the program of the meeting, I had noticed that Paul Kelly also was an invited speaker. So, I happily thought, I will finally meet Paul Kelly. While I was at the registration desk in Corpus Christi, a man advanced toward me with an extended hand. As we shook hands, he said, "I'm Kelly." "Ah," I replied, "the great geometer." "Oh, no," he said, "I'm not Leroy Kelly, I'm Paul Kelly."

Perhaps this is a good place to tell an anecdote about each of the two Kellys.

Leroy Kelly, when teaching a freshman course in analytic geometry at the University of Missouri in about 1946, once casually remarked in his class that if the inside of a race track is elliptical, and the track is of constant width, then the outside of the track is not necessarily an ellipse. Among the students in the class was one truly "from Missouri," who asked to be shown that the outside need not be an ellipse. Now this was not the easiest thing to do at the student's level of mathematical understanding. As a result, Professor Kelly proposed, as Problem E 753 in the January 1947 issue of *The American Mathematical Monthly,* the question: "How can one convince a class in elementary analytic geometry that if the inside of a race track is a noncircular ellipse, and the track is of constant width, then the outside is not an ellipse?" Solutions, some of which may have been beyond the student, appeared in the September 1947 issue of the journal.

Paul Kelly experimented during the 1969 fall semester with a mathematics appreciation course at the University of California in Santa Barbara. The course was open to the general college student, and it was Professor Kelly's hope in the course to relate various mathematical ideas to the real world of the student. For example, he thought it would be interesting to the student to note a possibly important implication of non-Euclidean geometry for understanding oneself. His thesis ran as follows:

From early school days we get used to the Euclidean explanation of space, and for ordinary problems this system of ideas works—indeed, it works so satisfactorily that we regard it as the "true" system, and we tend to feel an ingrained disbelief in non-Euclidean geometry. Now each person is a kind of system of beliefs, emotions, attitudes, and so forth, picked up from his parents, teachers, society, and experiences in general, and for him, in ordinary circumstances, this system "works." As a result he tends to regard his system as "true," and to view with suspicion, perhaps even with contempt or disbelief, any personal philosophy, religion, or culture that differs from his own, such as those of foreigners or of people in a different stratum of society. Achieving a genuine understanding of someone radically different from oneself is thus seen to be emotionally parallel to achieving a genuine understanding of the possibility of our world being closer to some non-Euclidean model than to a Euclidean one.

Some Debts

A Pair of Old Debts

I've lectured in every state of the union except Alaska, and many times in several states. Another frequent lecturer is Father Stanley J. Bezuszka, of Boston College. Father Bezuszka is a little human dynamo, and it was natural that we would occasionally appear on the same speaking program. It was in this way that we got to know one another.

Thus, during an intermission of an NCTM meeting held in Corpus Christi some years ago, I found myself at a little table enjoying a Coke with Father Bezuszka. The conversation turned to geometry and the old master, Euclid. I finally said, "I owe an immeasurable debt to Euclid. Reading the first six books of his *Elements* in school marked an important turning point in my life, for it determined that I would go into mathematics as my life's work." And I concluded by musing, "Isn't it remarkable that what a man did some 2000 years ago should so affect one's life?" And then I suddenly realized that perhaps there was nothing so remarkable about it after all, for there sitting across the table from me was Father Bezuszka in clerical dress, and he similarly had his life markedly affected by the doings of a man who lived about 2000 years ago.

To Dr. William D. Taylor

One of my earliest teaching appointments was that of Assistant Professor of Applied Mathematics at Syracuse University. In addition to teaching various courses in engineering mathematics, I was asked to teach a course in mechanics. Since I had had little experience with mechanics, I decided to help prepare myself by sitting in on the elegant lectures on the subject given at Syracuse by the lovable Professor William D. Taylor (1870–1945). I owe a great deal to those lectures.

I shall never forget the opening class meeting. Professor Taylor, who was a deeply religious man and a former Episcopalian minister, commenced the course with a solemn prayer in which he asked the Lord to help his students master the beautiful subject of mechanics, so that they might emerge from the course with good grades and an abiding appreciation of the material. Though it turned out that the first wish was not granted to every student, surely the second one was.

My personal innovation in the Mechanics course was to teach it via the axiomatic method. With two assumed principles one can obtain all of Statics, and with one more assumed principle one can obtain all of Dynamics. The three assumed principles are easily made creditable by simple experiments performed at the lecture desk.

Hypnotic Powers

In 1943, while teaching Mechanics at Syracuse University, I discovered that I possess hypnotic powers. In the class was a lovable, but mentally poorly equipped, student who went, among his classmates, by the simple name of Snuffy. One day, early in the semester, I tried to make clear the difference between conclusions arrived at by deduction and those arrived at by induction—the former (if the premises are accepted) being incontestable, while the latter are only more or less probable.

I illustrated deductive reasoning by such examples as:

Premises:

(1) All Canadians are North Americans.

(2) All Nova Scotians are Canadians,

Conclusion: All Nova Scotians are North Americans.

To illustrate induction I said, "Suppose I have a bucket of pebbles here on my desk. I select a pebble from the bucket," I pantomimed picking a pebble from my imaginary bucket, "hold it over the edge of the desk," I moved my hand containing the imaginary pebble beyond the edge of the desk, "and release it." I opened my hand. "I note that the pebble falls to the floor." With my eyes I followed the descending pebble as it fell. "I try the same experiment, and again the selected pebble falls to the floor. I do this with 100 pebbles, and in each case the pebble falls to the floor. Interesting! I grab a notebook and record my discovery: *If a pebble is released, it falls to the floor*. Now," I continued, "I haven't really proved my conclusion. I merely have overwhelming probability in favor of the conclusion. For all I know, if I should select a 101st pebble from the bucket and release it over the edge of the desk, it may rise to the ceiling rather than fall to the floor." As I was making these statements, I pantomimed my remarks, finally following the 101st pebble with my eyes up toward the ceiling of the classroom. I noticed that Snuffy had followed the imaginary 101st pebble upward, and he sat for some time gazing at the ceiling. Finally a fellow student gently slapped him, and as he came to he asked, "Professor Eves, what made that one go up?"

Poor Snuffy did very poorly in Mechanics, but I was saved from flunking him, for before the course was over Snuffy withdrew from college and joined the army. We later heard that he took part in the African campaign. General Rommel had been pushing the Allies all over North Africa, but as soon as Snuffy arrived the tide turned and the Allies became successful. So perhaps Snuffy *did* have some remarkable ability. But, then, this conclusion has been arrived at by induction.

At any rate, the moral of this tale is: When I'm lecturing be very careful not to let me look you straight in the eye.

Founding the Echols Mathematics Club

It was, if I recall correctly, during my junior year (1933) at the University of Virginia that a group of us mathematics students decided we would like to establish a university mathematics club. It wasn't long before a charter was drawn up and an appropriate schedule of times of meeting decided upon. There only remained the matter of naming the club. I suggested we call it The Sylvester Mathematics Club, basing my suggestion on the following bit of early history of the university.

In 1841 J. J. Sylvester accepted an appointment of Professor of Mathematics at the University of Virginia. He accordingly came to America and entered into his new duties with all the enthusiasm and energy of his youthful twenty-seven years. But it soon became apparent that there were students who resented the presence of a foreigner and a Jew on the faculty, and Professor Sylvester began to suffer annoyances in the classroom. Finally, after three months of growing harassment, Professor Sylvester reported a case of serious disrespect accorded him in the classroom by a Mr. W. H. Ballard. Mr. Ballard was summoned to tell, in the absence of Professor Sylvester, his side of the altercation. His report, of course, was at variance with that of Professor Sylvester, and he had his view of the affair backed up by his crony Mr. W. F. Weeks. Upon hearing of the introduction of this biased witness, Professor Sylvester protested, and wondered why some more neutral student witness had not been employed instead. In an attempt to carry out Professor Sylvester's wishes such witnesses were next called, but they felt they should support Mr. Ballard's report. At this point, the faculty records become obscure, but it seems that finding himself in discord with both the students and his fellow faculty members—hurt and dissatisfied with the lack of a firm faculty decision—Professor Sylvester submitted his unconditional resignation. The resignation was accepted and Sylvester left Virginia. He went to New York City, where he had a brother living, and tried unsuccessfully for about a year to find other gainful employment in America. Failing, he returned, penniless, to England.

Fortunately for America, in 1876 Sylvester once again crossed the Atlantic, this time to take a position at the newly founded Johns Hopkins University in Baltimore. Sylvester remained at Johns Hopkins for seven years, and those seven years proved to be among the happiest and most productive of his life. It was during his stay at Johns Hopkins that Sylvester founded, in 1878, the *American Journal of Mathematics*.

Since Sylvester was the first truly top-notch mathematician to teach at the University of Virginia, the selection of his name was natural. Moreover, the naming of the club after him could be regarded, though rather late, as a slight remedying of the former injustice done him.

But the mathematics club was not fated to be named after Sylvester. For there was on the mathematics faculty at the time a lovable old professor—a fine mathematician and a good instructor—Professor William Holding Echols, who was just about to retire. After all, we reasoned, Professor Sylvester received many honors during his lifetime; therefore let us in our small way honor Professor Echols by naming the club after him,

Professor Echols was a tall, gaunt, spectral figure, who looked as though he had just risen from a coffin. He taught the calculus course, using the famous Granville text, and all of us in our little group had studied the subject under him. It was difficult asking him questions, as he was almost stone deaf. It was always amusing to watch him erase his blackboard with a sleeve of his jacket.

J. J. Sylvester

I fondly recall Professor Echols. His teaching procedure was first to present the new material on the front blackboard, and then to send the class to the blackboard around the sides of the room. Problems would be assigned to us and he would slowly rotate about the room observing how we were doing, while his able assistant Tom Wade would rotate in the reverse direction. One day, as he stood behind me while I was working on the board, he pointed a long bony finger at a digit I had written in my work and gruffly asked:

"What is that?"

"It's a four," I replied.

"What?"

"*It's a four,*" I replied more loudly.

"What did you say?"

"IT'S A FOUR," I replied even more loudly.

"What?"

"IT'S A FOUR," I yelled.

"Only you would know that," he said, and he passed on to the next student.

Professor Echols always appeared in a rather rumpled looking condition, in an unpressed suit and his chalk-covered sleeve. He always kept a window open on the side wall at the end of the front of the classroom. Each morning, about half way through his introductory lecture, he would give a great hack in his throat, trot over to the open window, and spit out a gob of annoying phlegm. One day some student, unknowingly to the professor, closed the window—with rather grisly results. It was not uncommon at the end of a class session, when we had all returned from the blackboard work to our seats and Professor Echols was at his board engaged in some last minute clarification of a persistent error he had observed in our work, that the bell marking the end of the period would ring. The class would quietly leave by the rear door while Professor Echols, unaware of the signaling bell, continued plodding along at his board with his back to an empty classroom.

Professor Echols had trouble remembering the names of his students. On one occasion he had two students in the same class named Smith and Jones, and he was always confusing them. At the end of the term, after grading papers, he met one of the two students, who inquired about his grade. Professor Echols replied, "If your name is Smith, you passed; if Jones, you failed."

All this was over sixty years ago. But the Echols Mathematics Club of the University of Virginia is still in existence today, holding regular meetings. I wonder how many of the present members of the club have any idea of just who the name Echols stands for.

I've been instrumental in founding a number of other student mathematics clubs, and of a few faculty mathematics seminars and colloquia. But it was my founding, very shortly after I joined the teaching staff at the University of Maine, of the Northeastern Section of the MAA that pleases me the most. This latter, however, requires a story all of its own.

Meeting Maurice Fréchet

It was in the mid-1940s that the great French mathematician Professor Maurice Fréchet toured up the west coast of the United States, visiting a number of institutions of higher learning on the way. When he reached Corvallis in Oregon, we members of the Mathematics Department at Oregon State College* had the pleasant opportunity of meeting the celebrated man. Professor Milne, the fatherly chairman of our department, hosted the event at his attractive home in the scenic countryside a couple of miles from Corvallis.

It was Professor Fréchet who, in 1906, inaugurated the study of abstract spaces, and very general geometries came into being that no longer necessarily fit into the former neat Kleinian definition. A space became merely a set of objects, usually called *points,* together with a set of relations in which these points are involved, and a geometry became simply the theory of such a space. A result was that a host of special abstract spaces were created, such as *Hausdorff spaces* and other more general *topological spaces*.

Professor Fréchet was a charming conversationalist, skilled in the English language, and we found it interesting comparing and contrasting university customs in our two countries. I recall how surprised Professor Fréchet was to learn that in our country (at that time) it was mandatory for professors to retire

---

*Now Oregon State University.

Maurice Fréchet

at age 65. Professor Fréchet had already entered into his 70s, still teaching and expecting to teach for many more years. He lived to the ripe age of 96; I wonder when it was that he finally left the classroom.

Some twenty years later, I once again made contact with the eminent French mathematician when I translated from French into English the little gem, *Initiation to Combinatorial Topology,* written by him and Ky Fan. It came about in this way.

Paul Prindle, Art Weber, and Bob Schmidt, three very able renegades from Allyn and Bacon, decided to found their own publishing company, and to devote it entirely to the publication of mathematics texts. Since I had written several mathematics books for Allyn and Bacon, the new group asked me to write a book for their newly-formed company. I had just finished the manuscript of my two-volume *Functions of a Complex Variable,* and offered it. The book was accepted and became the first PWS book. I was invited to accept the position of Editor of the Complementary Series, a series of paperbacks which the company planned to start off with. My book appeared in 1966, as the first volume of the series. I rapidly introduced further books into the series, and chose the little book by Maurice Fréchet and Ky Fan as the seventh one. It came out the following year, in 1967.

I might append a couple of comments connected with my *Functions of a Complex Variable* text. It was the only one of my books that was submitted for publication in hand-written script, rather than in typed form. The work was soon translated into Spanish and, in this guise, was adopted by the University of Mexico. Shortly after the adoption, some sort of student protest broke out at the University of Mexico, and some unkind individual suggested that it was probably the adoption of my text that fomented the unrest.

Mathematizing the New Mathematics Building

When I first arrived on the University of Maine campus, the Mathematics Department Office was housed in Stevens Hall along with a few mathematics classrooms, and offices and classrooms of other disciplines. The rest of the mathematics classrooms were scattered in various buildings around the campus, some even in a temporary wooden structure.

Then a large, new, four-story building was constructed, to be shared by the English and Mathematics Departments. The building ran roughly north and south, with foyers and two auditoriums at the north end. The rest of the building was devoted to classrooms and offices, with the northern half the domain of the English Department and the southern half that of the Mathematics Department. Attached exteriorly to the south end of the building was a tall stairwell mounting up all four floors of the building and presenting a long vertical unadorned brick wall.

A request circulated seeking suggestions to characterize each half of the building. I reacted with enthusiasm and soon drew up a list of a number of ideas for the mathematics end. Here are a few samples of my suggestions.

1. The room numbers. It would be interesting and instructive to accompany each familiar Hindu-Arabic room number with the same number expressed in some other historic numeral system. Thus we might have, for example:

Room 311

Room 312

Room 313

Room 314 CCCXIV

Room 315

Room 316

Room 317

Room 318

It would be particularly fitting in an institution of learning to have such a blending of new and old cultures.

2. *The hallway walls.* The digits in the decimal expansion of π seem to satisfy no pattern and appear to occur in an absolutely random fashion. Because of this, the successive digits in the decimal expansion of π have been used where a random sequence of digits is required. Use of this randomness can be made in laying tile, where, to break the monotony of a uniform color, one wishes every here and there to insert a tile of a different color.

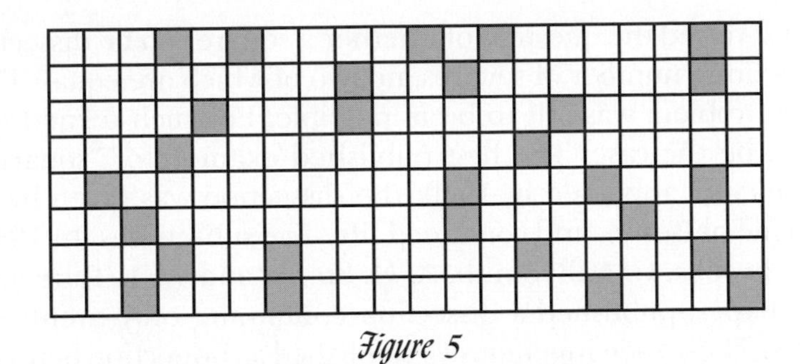

Figure 5

How very interesting it would be if panels of the lower parts of the hallway walls of the new mathematics building were brightened by tiles in this "random" fashion, one panel utilizing the decimal expansion of π, another that of $\sqrt{2}$, still another that of e, and so on. A panel of this sort, employing the decimal expansion of π is shown in Figure 5.

3. The foyer floors. In 1925, Z. Morón noted that a 32×33 rectangle can be dissected into nine squares, no two of which are equal. See Figure 6.

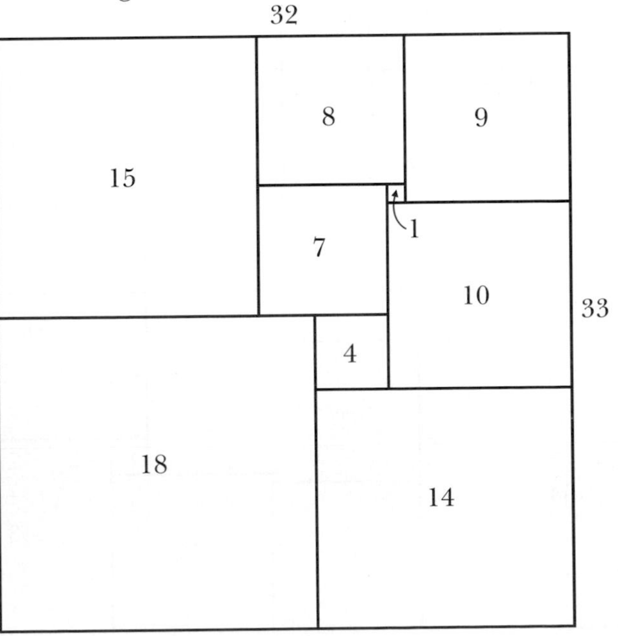

Figure 6

This raised the question of whether a square can be dissected into a finite number of squares, no two of which are equal. This latter problem was felt to be impossible, but such turned out not to be the case. The first published example of "squaring the square" appeared in 1939; the dissection was given by R. Sprague of Berlin, and contained fifty-five subsquares. In 1941, R. C. Brooks, C. A. B. Smith, A. H. Stone, and W. T. Tutte, in a joint paper, published a dissection containing only twenty-six pieces. These men ingeniously established a connection between the dissection problem and certain properties of currents in electrical networks. In 1948, T. H. Willcocks published the twenty-four piece dissection of a square into unequal subsquares pictured in Figure 7, and this dissection is today the record so

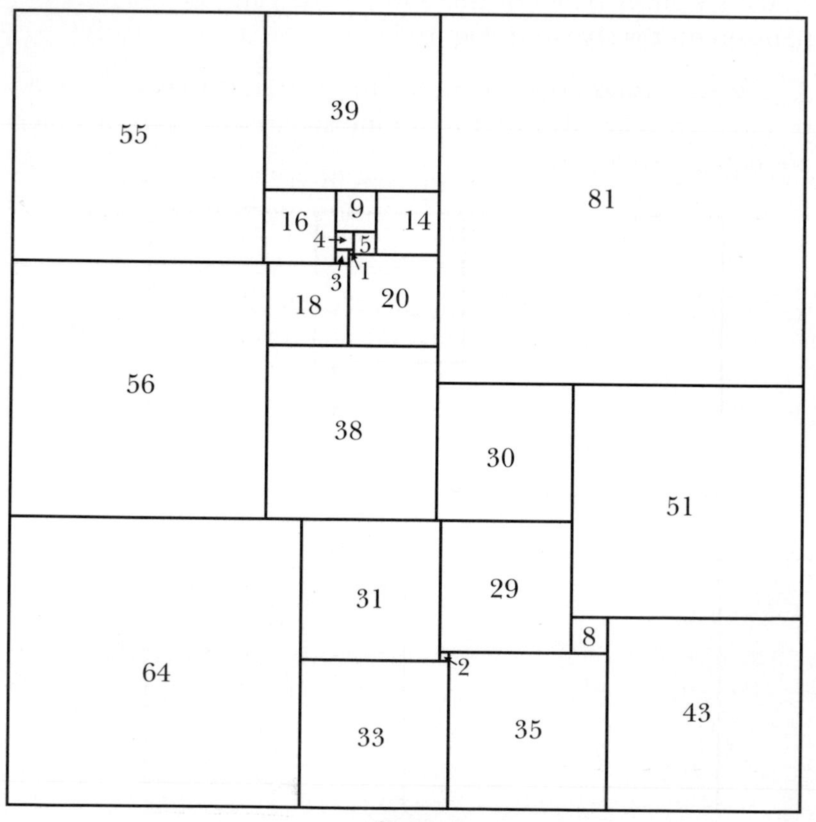

Figure 7

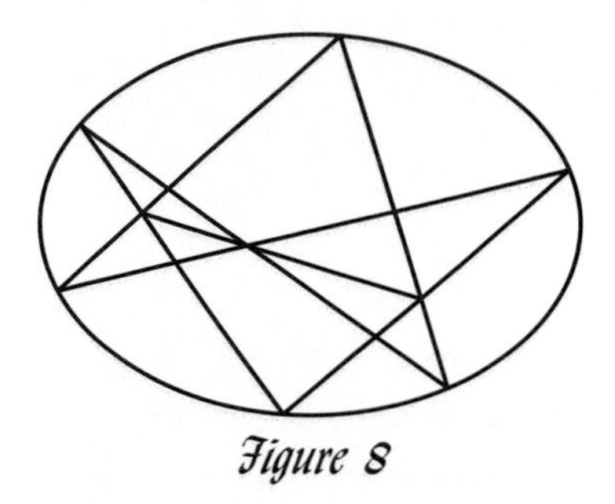

Figure 8

far as the least number of pieces is concerned. Surely, somewhere in the planning and designing of a mathematics building at a university, the Willcocks squaring should appear. And what better place than in the floor of a foyer to an auditorium.

There is another excellent candidate for a foyer floor. In projective geometry there are few prettier or more fertile theorems than the famous *Pascal mystic hexagram theorem,* which asserts that "the three points of intersection of the three pairs of opposite sides of a hexagon inscribed in a conic section are collinear"—see Figure 8. Here then is another fine mathematical pattern for the floor of a foyer, using, perhaps, inlaid brass bands for the lines of the figure.

4. The southern stairwell. How better to distinguish the new mathematics building from all other buildings on the campus than to run up and down, on the exterior of the long vertical southern stairwell, an enormous stainless steel integral sign. The symbol would be visible from many parts of the campus.

None of my suggestions for mathematizing the new mathematics building was adopted, the cost factor being considered as prohibitive. Some years later, however, Clayton Dodge did manage to incorporate π on one of the hallway walls.

It was John Mairhuber, I believe, who came up with the perfect name for the new English-Mathematics Building—Lewis Carroll Hall. Unfortunately, this suggestion, too, was not adopted. The building is today known as Neville Hall, named after a former president of the university.

Finding Some Lost Property Corners

When I began seeking a teaching position the great depression was at its worst, and openings in teaching were almost nonexistent. To mark time until the situation might better itself I found a job assisting a surveyor. The following year I took examinations and became a licensed land surveyor for the state of New Jersey. Rather than compete with the established city surveyors, I set up practice in the country and performed farm and vacation plot surveys. Although the remuneration was slight, it was a healthful and enjoyable outdoor activity. There was one farm survey I did for a hundred pounds of potatoes. Potatoes at that time were a penny a pound, so I made a dollar for my work—not much moneywise, but a hundred pounds of potatoes constituted a lot of food. Later I worked as a surveyor for the General Housing Corporation of New Jersey, and received the munificent recompense of 25 cents an hour. An odd number of situations occurred in my survey work. Here is an account of two of them.

A man with a small piece of property along a country road lost the position of one of his front corner points. He had put in a driveway leading in a slant from his garage out to the road, and wanted to be sure it lay entirely on his property. Being located down a little country lane necessitated starting from an established bench mark over a quarter of a mile away and then following a crooked course to his corner. I finally arrived at a point along the road edge in the middle of the end of his driveway. I drove a large spike in the ground to hold the point, and called the property owner. I told him that his deed called for a driven hollow iron pipe at that corner, and I asked him if he had ever seen a hollow iron pipe driven into the ground somewhere in the area. "Yes," he said. "There was a pipe driven in the ground where I was putting in my driveway. I couldn't pull the pipe out so I drove it deeper into the ground. Was that pipe my corner?" "Probably so," I replied. I then dug into his driveway around my spike, and lo, there was my spike neatly driven into a hollow iron pipe. It sure made my survey work look amazingly accurate. But there must have been an element of luck in it, for the probability of arriving at the precise spot by a polygonal chaining starting over a quarter of a mile away must be very small indeed. It did prove, though, that it is wiser to have your property surveyed before putting in a driveway rather than the other way about.

In another instance there were two farmers who disagreed as to the position of a point separating their two properties. The deeds called for an oak stake, but time had completely obliterated it. After performing a survey leading to the area of the lost point, I marked the place and began a careful excavation. I soon came upon a small square of discolored earth. Digging deeper, the discolored square continued downward for about a foot and then grew smaller and smaller until it disappeared. I extracted some of the discolored material, put it in an envelope, and had it chemically analyzed. The report came back that the material was rotted oak wood. Thus the lost corner was restored.

Surveying is a practical exercise in trigonometry—plane trigonometry if the survey covers a small area, spherical trigonometry if the area is so large that the curvature of the earth must be considered. Several eminent mathematicians employed their mathematical expertise as surveyors. For example, Thomas Harriot (1560–1621), known as the founder of the English school of algebraists, was sent in 1585 by Sir Walter Raleigh as a surveyor with Sir Richard Granville's expedition to the New World to map what was then called Virginia but is now North

Carolina. And Carl Friedrich Gauss (1777–1855) carried out a geodetic triangulation of Hanover, and measured a meridional arc. Some other mathematicians associated with surveying are: Gerhardus Mercator (1512–1594), Giovanni Domenico Cassini (1625–1712), Pierre Louis Moreau de Maupertuis (1698–1759), Joseph Louis Lagrange (1736–1813), and Gaspard Monge (1746–1818).

I once experimented teaching trigonometry by augmenting the classroom work with surveying work performed on the university campus. The class greatly enjoyed seeing trigonometry actually employed in finding large polygonal areas, the heights of tall buildings, the width of a large river, and so on. But it did demand more time from the students. I also think I may have been the first in the United States to teach a course in trigonometry with the aid of calculators. This was in the days before computers and we used small inexpensive hand-cranked Monroe calculators bolted to a desk in the calculating lab. Of course, the slide rule had been used by many instructors. Logarithms, as a calculating device, have become obsolete.

The Tennessee Valley Authority

In 1941 the Tennessee Valley Authority in Chattanooga hired me as one of a group of mathematicians to convert logarithmic surveying forms to calculating machine surveying forms. It was the heyday of the calculating machines—the Marchant, the Frieden, and the Monroe. These machines were about the size and heft of the then standard typewriters. TVA chose to use Monroe machines.

The forms had to be arranged so that once the given information was inserted, any person familiar with a calculating machine could step by step arrive at the desired final result. That is, the forms had to be self-explanatory. They were also to contain a built-in check of the calculation work. The people doing the calculating didn't have to know anything about surveying. It was felt that if for a particular problem three forms should be passed out in the calculating office, and if on return all three self-checked and agreed with one another, then the final result could be unhesitatingly accepted.

Consider, for example, the important *three-point problem.* At that time all states of the country, and many important foreign countries, were on coordinate systems. That is, each point of a concerned region bore coordinates with respect to some appropriate frame of reference. Suppose there are three points A, B, C with known coordinates. A transit is set up at a point P of unknown coordinates and angles APC and BPC are read. From this information, along with trigonometric tables, one is to find the coordinates of the point *P.* Forms were to be designed so as to minimize the number of entries taken from tables. The form that I devised for TVA requires only two such extractions, namely the cotangents of the two measured angles. The form was later published in *Civil Engineering* (Jan. 1945, p. 33), and then was adopted for survey work by the government of Argentina.

An important project of the TVA at the time I was there was the strategic military mapping of the St. Lawrence Valley. Britain was undergoing a pounding German blitz. Should Britain fall to the Germans, the military strategists reasoned that Canada might then also fall. With Canada under German dominion, the United States would be threatened. The logical invasion of the United States from Canada would be via the St. Lawrence Valley. To prepare for such an eventuality it was considered important to have reliable military maps of the area. So survey crews were sent to do this mapping.

Now in surveying a point is frequently reached where, in order to proceed, the surveyor needs some length or angle calculable from his earlier fieldwork. Thus a surveyor's job consists of an alternating sequence of fieldwork and calculating work. The usual surveyor does both of these things himself, keeping the field and calculated information recorded in a *field book.* This was not the way TVA operated. Their surveyors' field books were loose-leaf. When a point was reached by a surveyor that required some calculation before proceeding, the concerned pages of the field book were extracted and mailed to the TVA calculating center in Chattanooga some thousand miles away. While the calculations were being made at Chattanooga, the surveyor up in the St. Lawrence Valley would proceed with some other project (of which there were many), and when the mailed-in pages returned with the required calculations, he would continue his former piece of work. A remarkable and unusual procedure, but quite successful on large surveying projects.

After I left TVA my next encounter with surveying was at Syracuse University. There I got interested in the then new procedure of mapping terrain from a series of photographs taken through an opening in the floor of an airplane. This type of surveying is known as *photogrammetry* and a journal called *Photogrammetric Engineering* came into existence to record papers and research on the subject. I published a number of papers in this journal. Since an airplane was not able to fly in a perfectly horizontal plane, to obtain true maps from the photographs, the photographs had to be corrected for tilt and swing. And of course there were many other interesting problems. The mathematics of photogrammetry consists largely of exercises in the analytic geometry of three-space.

Back in those days, no one envisaged the remarkable computers to come into existence within a relatively few years. The result is that all that early work at TVA and at Syracuse is now largely merely an interesting museum piece. This has been the fate of most of my work in applied mathematics. Nevertheless, I often look back upon those challenging days with considerable pleasure.

A piece of surveying work that I did at Syracuse, and that may never become obsolete, concerns my ideal highway transition spiral. But the story of this would require an account all to itself.

How I First Met Dr. Einstein

I hadn't been on the Princeton campus but a few days when, one morning, while walking along a street of the village, I saw Dr. Einstein plodding along in the opposite direction on the other side of the street. I had forgotten that this was the year he was to assume his life-time appointment at the Institute for Advanced Study—that he and I both would be "freshmen" together on the Princeton campus. I paused to watch the great scientist pass by. While so engaged I noticed a young fellow on the other side of the street walking toward Dr. Einstein. When he was about fifty feet from Dr. Einstein he looked up and saw the famous man approaching him. He stopped, and then suddenly darted across the street to where I was standing. I saw Dr. Einstein give a little puzzled shrug.

"Why did you suddenly cross the street when you saw Dr. Einstein approaching you?" I asked my new companion.

"Oh," he replied, "I thought I should, otherwise I might disturb him while he was contemplating some deep theory."

On an afternoon a couple of days later, I was again walking along the same street, when I noticed Dr. Einstein plodding along ahead of me in the same direction. I increased my pace to catch up with him, and when I reached him I asked, "May I walk along with you, Dr. Einstein?"

"Oh, I wish you would," he replied. "It seems no one will talk to me or walk with me."

I hastened to explain that it wasn't because they didn't want to, but that they were afraid of distracting him should he be contemplating some deep theory in physics. This caused him to stop, and turning to me he said, "Ach, but dat ist nonsense."

During our stroll together I found that he was living on Mercer Street, just a few blocks beyond the dormitory in which I was housed. He said he enjoyed Princeton, and liked to take early morning walks about the streets of the village. I told him I liked doing the same thing. The outcome was that I should call on him at his home on pleasant early mornings, so that we could ramble together. Thus started our early morning walks together, during which we discussed all sorts of interesting things.

Adolf Hurwitz (center) with Albert Einstein and Hurwitz's daughter, Lisi

Catching Vibes, and Kindred Matters

One winter day at Princeton, after a light snowfall, I accompanied Dr. Einstein over to Fine Hall. As we were walking along, I sensed that we were being followed, so I turned my head and looked back. There, about a dozen paces behind us, I saw a freshman physics student, whom I knew, carefully putting his feet one after the other in Dr. Einstein's footprints. He did this for about half of a block. The next day I met the student and asked him why the day before he had walked in Dr. Einstein's footprints.

"Oh," he said. "I had a tough physics test coming up that morning, and I thought that if I walked in Dr. Einstein's footprints I might perhaps catch some useful vibes."

"Did it work?" I asked.

"No, not at all," he mournfully replied.

"Why didn't you walk in my footprints?" I asked.

He looked at me somewhat startled and unkindly said, "Do you think I'm *that* crazy?"

So it would seem that there is not much in such ideas as catching vibes, or in numerological and astrological nonsense. Consider, for example, the following facts. Garrett Birkhoff and I were fellow graduate mathematics students at Harvard. I learned that we were both born on the same day, of the same month, of the same year, in the same state of the union, and in towns with names starting and ending with the same letters. Now, by the stars, we should have turned out about equally well. But Garrett became an outstanding mathematician, whereas I remained only so-so. It follows that there just can't be much in similar astrological data.

"But ah," someone once remarked, "you didn't consider the important fact that you were born in *Paterson,* New Jersey, whereas Garrett was born in *Princeton,* New Jersey.

A Pair of Unusual Walking Sticks

One morning at Princeton, as I was walking over to Dr. Einstein's home to pick him up for an early morning ramble, I passed a curbside pile of brush awaiting the town's disposal truck. The brush consisted largely of curved canes about four feet long, an inch thick at one end and tapering to about a quarter of an inch thick at the other end. I selected one of the canes and found that if I held the thick end in my hand, extended the cane (arched upward) down to the sidewalk in front of me, and walked forward, the tapered end would chatter along the sidewalk. Of course, for it to work, one had to walk it along a hard surface, like a cement sidewalk or a macadam street.

Equipped with my stick, I called at Dr. Einstein's home on Mercer Street. As we set out for our walk, Dr. Einstein was greatly intrigued by the behavior of my walking stick. "I wish I had one of those," he said. "Here, take mine," I replied. "Oh, I couldn't take your stick," he said. "Well," I then said, "I know where we can get another like it." "Let's do that," he replied. So we went to the brush pile and selected another walking stick. With the two sticks preceding us, and Dr. Einstein beaming like a little boy, we continued our early morning ramble.

When on subsequent mornings I called at Mercer Street, Dr. Einstein would ask, "Did you bring your stick?" If I did he would ask me to wait a moment while he skipped upstairs to get his stick. I have often wondered what early morning folks thought when they saw the great scientist and a young graduate student walking along, side by side, pushing their curious sticks before them.

When the time came for me to leave Princeton, I placed my walking stick on the mantle over the fireplace of my room, with a detailed note describing how to use it. I wonder if my successor ever took advantage of the stick's comforting chatter when he went awalking. I have also wondered what ultimately became of Dr. Einstein's walking stick. In any event, I dare say it's quite certain that the two sticks never again enjoyed a walk together.

A New Definition

I have told over sixty Einstein anecdotes in my various Circle books, but only those in which I was personally involved can properly be considered as reminiscences. Here are three that, when I told them in my Circle Books, I felt too forward to mention my small role in them. But since they all evoke pleasant personal memories I will here retell them in more complete form. The first two illustrate Dr. Einstein's sense of humor.

One day I accompanied Dr. Einstein to a lecture by a visiting physicist. The lecture was extremely dull, and the speaker droned on and on and on. Finally, Dr. Einstein cupped his hand about one side of his mouth, leaned close to my ear, and whispered, "I now have a new definition of infinity."

Mystery

Dr. Einstein and his wife were invited to a banquet, in honor of the great scientist, given at the Waldorf Hotel in New York City. Mrs. Einstein was thrilled and purchased a beautiful new gown for the occasion. But she unfortunately developed a very bad cold, and found she could not attend. It was then that I was asked to substitute for her, to see that Dr. Einstein got safely to New York and home again. It was a formal affair, with the men in white ties and the ladies in elegant décolleté evening gowns.

When, after the affair, we arrived back in Princeton, we found Mrs. Einstein waiting up for us, eager to learn what had taken place. Dr. Einstein began to tell her about the famous scientists who were present, and the fine things they said about him, but she cut him short.

"Never mind all that," she said. "How were the ladies dressed?"

"I don't know," replied Dr. Einstein.

"Now, Albert, you have eyes in your head. How were the ladies dressed?"

"I really don't know. Above the table they had nothing on, and under the table I was afraid to look."

Dr. Einstein's First Public Address at Princeton

There was a small auditorium, that I estimated would hold about 200 people, located in the center of the ground floor of Fine Hall (Princeton University's mathematics building). It was furnished with comfortable fold-up theater seats and a small stage with a blackboard. I was informed that the auditorium was used by occasional visiting scholars to give lectures to interested brethren. It seemed to me, with the Institute for Advanced Study based at the University (the Institute did not have its own buildings at that time), the little auditorium would be an ideal place for regular bi-weekly meetings at which Institute members, and others, could describe some of the work which engaged them. The idea took hold, but, like such ideas, where the innovator is catapaulted into the chairmanship of the new concept, I found myself saddled with the task of securing the first speaker.

Thinking the matter over it occurred to me that Dr. Einstein, fresh on the Princeton campus, would be an ideal first speaker. I accordingly broached the matter to the great scientist. After a bit of squirming, he said, "I don't know anything those people wouldn't already know." "Would you consider a suggestion?" I asked him. "Oh, yes," he replied, "If you can think of something suitable I would certainly consider it."

Now the reason I was at Princeton was that at Harvard I had specialized in differential geometry, under Professor Graustein, a fine geometer and chairman of the Harvard Mathematics Department. We chiefly engaged ourselves in vectorizing, and thus condensing, the subject. It was rumored that there was an even greater condensing tool, called tensor analysis. Though, at the time, tensor analysis enjoyed a European vogue, there were few mathematicians in America versed in the new technique, none at Harvard. The chief center of tensor analysis in America was Princeton University. Professor Graustein accordingly wangled an appointment for me at Princeton, where I was to study the tensor methods and then bring them back with me to Harvard. Now Dr. Einstein employed tensor procedures in his work, and had even contributed to the subject. So I asked Dr. Einstein:

"Would you be willing to discuss your contributions to tensor analysis?"

"If you think it would interest the attendants, I am willing to try," he replied.

And so the first talk of our select group was settled, and I placed a small 3x5 card on the Fine Hall bulletin board, announcing the speaker, the topic, and the time and place of the meeting.

On the day of the meeting, I picked up Dr. Einstein to walk him over to Fine Hall. A surprising sight met our eyes. The campus was crowded with cars, many even parked on the University's lawns. One saw such a sight only on the occasion of a Princeton-Yale football game, or some other big athletic event. As we approached Fine Hall, the crowd of people became denser and denser, and inside the Hall there was a milling mass trying to get into the small auditorium. Apparently my little 3x5 card on the bulletin board had been noticed by students, and thence spread to the Princeton townfolk. Students wrote home to parents, who arrived, along with friends that they picked up. Apparently, unknowingly, I had invited Dr. Einstein to give his first public address at Princeton, and a large part of the world wanted to hear him.

With great effort Dr. Einstein and I managed to make our way through the crowd to the two seats, in the middle of the first row of the auditorium, that I had had the foresight to rope up as reserved for Dr. Einstein and myself, and from which, at the proper moment, I was to introduce the illustrious speaker. As we sat there in those two seats, Dr. Einstein craned his head over his shoulder looking at the pushing crowd. He then turned to me and said:

"Mr. Eves, I hadn't realized that in America there is so much interest in tensor analysis."

Parting Advice

As narrated elsewhere in these notes, after leaving Princeton University I marked time as a surveyor while waiting for a teaching position to open up for me. I finally secured a one-year appointment at Bethany College in West Virginia, filling in while the regular man was on leave working on his Ph.D. at the University of Chicago.

In the fall of 1938, before leaving for West Virginia, I decided to make a short nostalgic last visit to Princeton, and while there perhaps briefly to see Dr. Einstein again. I was lucky, for I found the scientist at home. He greeted me warmly and suggested we take a short stroll together. After walking a bit, he asked me how life had been since I left Princeton. I told him about my difficulty in securing a teaching position, and how I had been spending my time as a surveyor. He asked me if I enjoyed the work, and I told him it had its pleasant features, but that it was in no way as pleasing as my life at Princeton had been. I bemoaned the lack of kindred minds to talk mathematics with. In short, I more or less griped about my situation.

"I understand," Dr. Einstein said. "Now before we part, there is something I want to show you."

We had been approaching the end of our little ramble and were walking down Mercer Street to his house. On the final block, when almost at his home, he stopped and pointed downward to the sidewalk. I followed his pointing finger, and there in a crack in the sidewalk was a small wild aster, with pretty violet petals and a bright yellow center. We stood there silently

Drawing of Einstein and me by my daughter, Cindy Eves-Thomas

contemplating the little plant. I wondered how it managed to grow in such an unfavorable place. How did it get enough water? How did it escape being trodden upon?

Dr. Einstein seemed to read the thoughts that were passing through my mind, for, still pointing at the little flower, he gently put a hand on my shoulder and in a soft voice said, "Bloom where you are planted." I nodded, showing that I got the lesson. And as I stayed there by the little aster, Dr. Einstein went on to his home, paused on his front steps, turned toward me, and waved a farewell. I waved back to him, and that was the last time I saw Dr. Einstein.

In some earlier walk along his street, Dr. Einstein must have noticed the struggling little flower and sensed the lesson it conveyed. I happened to be the one in need of that lesson. At subsequent unfortunate junctions of my life I have tried not to gripe too much about things, but rather have tried to "bloom where I was planted." And I have, on occasion, passed the advice on to some students of mine.

Mythology of the Magic Square

The magic number of the square is 21. If you add the 2 and the 1 you get 3, which was the number assigned to Neptune of Roman mythology. Neptune was the god of the sea and the protector against storms and lightning. It was for this reason that sailing vessels of yore frequently painted the magic square on a deck of the ship—so that the ship would be safe from storms and lightning. A replica of the magic square on a barn or a house was considered better protection from lightning than the installation of lightning rods.

This magic square hangs on our barn in Lubec, Maine.

Two Newspaper Items and a Phone Call

The Mathematician and the Fundamentalist

At one time I lived directly across the street from a church officiated over by a fundamentalist minister who became concerned about my life in the hereafter. He told me grim stories about the lower regions and described the torrid horrors I might expect if I didn't snap to and change my ways by joining his church. One regretful day, in a jest that he completely failed to appreciate, I told him that I had no fear of the lower regions, because all the mathematicians and engineers who have gone there have undoubtedly remarkably improved the place with air conditioning and refrigeration.

Some time later I received the following item clipped from the Bangor *Daily News*, Thursday, January 8, 1970.

> HAD TO HAPPEN
>
> HELL, Norway (UPI)—The water froze in Hell Wednesday when the temperature dropped to 6 degrees below zero.

Too many engineers, I suppose, fooling around with the air conditioning. I do wish they would be more careful; they might spoil a good thing.

Moving to Gorham

Arrangements were made for me to spend a year or two assisting the Mathematics Department of Gorham State College, one of the units of the newly formed all-state University of Maine. Soon after the completion of the arrangements, a short two-column item, released by the public relations office of the University of Maine, appeared in the Bangor *Daily News*. It contained a picture of me with a brief statement that I and my family were temporarily moving from Orono to Gorham. The little item was boxed on the top and right side by a much longer, four-column communication, bearing a large headline that ran across all four co1umns that boldly asserted: NEW HOME FOUND FOR THE INSANE.

Wrong Number

There was a kindly old doctor, Dr. Asa Adams, in Orono, Maine, who looked after the health needs of many of the town's inhabitants, and who was often asked to be present, with his black medical kit, at many of the public affairs of the University of Maine—just in case his services might suddenly be needed there.

Our home in Lubec. The house is 150 years old.

One evening, while I was resting at my home in nearby Stillwater, the phone rang. I picked up the receiver and said, "Hello." A woman's voice at the other end of the wire asked, "Is this Dr. Adams?" "No," I replied, "this is Dr. Eves." There followed a lengthy silence, finally broken by the woman, who exclaimed in an exasperated voice, "Oh, come now!" And she hung up her receiver on its hook so angrily that it nearly shattered my eardrum. I sat for some time wondering what was the matter with the poor woman, and it was only considerably later that the reason for her strange action dawned on me.

Wherein the Author is Beasted

In one of my history of mathematics classes at the University of Maine I lectured on *gematria* or *arithmology*. I explained that since many of the ancient numeral systems were alphabetical systems, it became natural to substitute number values for the letters in a name. It was this that led to the mystic pseudo-science known as gematria, or arithmology, which became very popular among the ancient Hebrews and others, and was later revived during the Middle Ages. Part of this later gematria was the art of beasting—that is, cunningly pinning onto a disliked individual the hateful number 666 of the "beast" mentioned in the *Book of Revelations*: "Let him that hath understanding count the number of the beast: for he is a man; and his number is six hundred three score and six."

I gave several examples, such as Michael Stifel's effort to beast Martin Luther. The reformation led to a flood of attempts to beast the Pope at Rome. With skill one can often manage to beast almost anyone.

There are three popular methods of assigning number values to letters, and they are referred to as Latin key, Greek key, and English key.

In using Latin key, one considers only the letters in a name that represent Roman numbers, always taking U as V. For example, in

SILVESTER SECUNDUS (Gerbert, who reigned as Pope

Sylvester II), we have,

$$I = 1,\ L = 50,\ V = 5,\ C = 100,\ U = 5,\ D = 500,\ U = 5,$$

and

$$1 + 50 + 5 + 100 + 5 + 500 + 5 = 666,$$

the number of the beast.

In using Greek key, one writes the name in corresponding Greek letters, and then assigns the number values that these letters represent in the Greek alphabetical number system. For example

GLADSTONE

becomes, in corresponding Greek letters,

$$\gamma\lambda\alpha\delta\sigma\tau o\nu\eta.$$

We then have, in the Greek alphabetical number system,

$$\gamma = 3, \lambda = 30, \alpha = 1, \delta = 4, \sigma = 200,$$

$$\tau = 300, o = 70, \nu = 50, \eta = 8.$$

$$3 + 30 + 1 + 4 + 200 + 300 + 70 + 50 + 8 = 666,$$

the number of the beast.

In using English key, one takes A as 1, B as 2, C as 3, ..., Z as 26. The day following my lecture to the history of mathematics class, John F. Bobalek, one of my top students, with a smile, laid the following on my desk, to be subjected to English key:

HOWARD W. EVES, A PROFESSOR OF MATHEMATICS AND DOCTOR OF PHILOSOPHY.

The Scholar's Creed

There were three of us in our sophomore year at the Eastside High School of Paterson, New Jersey, who got to be known as the *three math nuts*. We were Al (Albert), Lou (Louis), and myself. We lived and breathed mathematics and were never happier than when we were struggling over some difficult mathematics problem.

We had been hearing a great deal about a remarkable branch of mathematics called "the calculus." Though today it is not uncommon to find a beginning course in calculus taught in high school, back then the practice was unheard of, and one had to await the sophomore year in college before being able to take the course. We just couldn't wait that long, and so we decided to learn the subject on our own. Now by far the most-used calculus book in those days was Granville's *Calculus*. What a mint of royalties that book must have earned, for it was the book almost universally adopted in the colleges throughout the country. We accordingly ordered three copies of the famous book and, at bi-weekly evening meetings at one another's home, we began teaching the subject to ourselves.

I know that today one is supposed to make disparaging remarks about the Granville calculus book, but I shall never be able to do this, for I admired and enjoyed the book enormously and learned a great deal from it. The problem lists were copious and challenging. Before we left high school we three had covered almost the entire book and had worked almost all the problems in it. When we later got to our sophomore year in college, lo and behold, the textbook for the calculus course was our old friend, the Granville text. No wonder we each soared through our college calculus course. There are few textbooks I have gotten to know so thoroughly, or which have given me so much enjoyment. My recollection of studying from it awakes such fond memories that I find it pleasant every now and then to get out my old copy of the book, reminiscently page through it and lovingly stroke its cover. Would I speak ill of the Granville text? Never.

We didn't always start right off with Granville at our evening gatherings. We might sometimes discuss a new approach one of us had come up with concerning a sticky geometry problem that had been eluding us, or discuss some other mathematical matter. This procedure almost always occurred when we met at Al's home, for he always served a round of lemonade or cocoa with cookies before we settled down.

It was at one of these sessions that I told the group of an interesting event I had read about in a remarkable book that I had acquired a couple of years earlier. The book was Florian Cajori's *A History of Mathematics*, and it was the second significant mathematics book that I had purchased. It was a comforting experience reading through Cajori's book, for I read about the history of so many fields of mathematics of which I knew essentially nothing, that I realized I had ahead of me a very delightful life attempting to master all these marvelous fields of study. It would take a long and pleasant life to assimilate all this material.

In reading the book I also came across many highly interesting little episodes in the lives of the great creators of mathematics. One of these episodes particularly seized my attention, and it was this event that I narrated at our little session at Al's home. The episode took place in the early part of the nineteenth century at Cambridge University in England, and involved three outstanding mathematics students there, who became intimate and lifelong friends. These three students met regularly on their own to discuss mathematics, and, after some months of these meetings, decided to subscribe to a vow. In great solemnity and earnestness, they collectively made the following pledge:

We vow to do our best to leave the world wiser than we found it.

And, as it turned out, these three men did precisely that.

Who were the three men? They were George Peacock, John Herschel, and Charles Babbage.

Peacock, born in 1791, was one year older than the other two. He was educated at Trinity College of Cambridge University, became Lowndean Professor there, and then Dean of Ely, a position which he held until his death in 1858. During his mathematical career he wrote his famous book, *Algebra,* in which he was the first to study seriously the fundamental principles of algebra and to recognize a purely symbolic algebra. This book, which occupies an important place in the history of mathematics, has in recent times been republished by the Chelsea Publishing Company. Peacock also, in his *Report on Recent Progress in Analysis*, inaugurated the series of valuable summaries of scientific progress that have continued to appear in the volumes of the British Association.

Sir John Frederick Herschel, born in 1792, was educated at St. John's in Cambridge. He wrote on the mathematical theory of sound. His discovery that sodium thiosulphate (commonly known today as hypo) dissolves silver salts gave photography an improved fixing agent. Later he traveled to Cape Town in southern Africa, and there surveyed and mapped the stars of the southern hemisphere. It was for this work in astronomy that he was knighted, as his father, William Herschel, had been knighted before him for his survey of the stars of the northern hemisphere. John Herschel died in 1871.

The third of the triumvirate of mathematics students at Cambridge University was Charles Babbage, who was born in the same year as his friend Herschel, and who died in the same year as did Herschel. Babbage was highly sociable, witty, gregarious, and the most colorful of the three students. It was in 1812 that he began work on his difference engine, to be succeeded about 1833 by work on his analytic engine. The work of Babbage on these "engines" provided the inspiration for the remarkable great mechanical and electronic computing machines that came into existence in recent years. Babbage had enunciated the principles on which all modern computing machines are based. When the British magazine *Nature* published an article in 1946 discussing one of America's first large calculators (the Harvard relay computer, Mark I), it entitled the article, "Babbage's Dream Comes True." In addition to his pioneering work on computing machinery, Babbage played a prominent role in the founding of the Astronomical Society (1812), The British Association for the Advancement of Science (1831), and The Statistical Society of London (1834). He also previsioned the present-day study called *operations research.*

Early in their student lives at Cambridge University, our three men founded the Analytic Society, the purpose of which was "to put English mathematicians on an equal basis with their Continental rivals." That is, the Analytic Society wished to remedy the severe situation into which English mathematicians had worked themselves following the earlier bitter controversy between Newton and Leibniz over priority of the creation of the calculus. The English mathematicians, backing Newton as their leader, had cut themselves off from Continental developments, with the result that mathematics in England suffered detrimentally for almost a hundred years. While on the Continent the mathematicians were using Leibniz's much more fluent differential notation dy/dx for the derivative, the English mathematicians were clinging to Newton's far less fortunate fluxional notation $\dot{y}$ for the derivative. In Babbage's humorous words, the Analytic Society advocated "the principles of pure *d*-ism as opposed to the dot-age of the university."

In 1816, Babbage, Herschel, and Peacock translated Lacroix's elegant one-volume *Calculus* into English. This did much to aid the Society's reform in the teaching and notation of the calculus. A further gain was made in 1817, when Peacock was appointed an examiner for the mathematical tripos; he replaced fluxional symbols by differential notation on the Cambridge examinations.

In short, our three men well lived up to their vow—they certainly left the world wiser than they found it. I was deeply impressed by their act of pledging to use their acquired knowledge in a manner that would benefit mankind, and it gave me pleasure reporting all the above to my two companions of the Paterson High School.

Preaching the baccalaureate (1977) on "The Scholar's Creed" at the University of Maine in Machias.

Al was a highly enthusiastic fellow, and when I finished narrating the above episode he exclaimed, "Why don't we three do the same thing?" We batted the idea around a bit, but feeling we had no corner on wisdom we doubted we would be able to leave the world wiser than we found it. By further wrestling with the matter, however, we came up with a somewhat different pledge, which finally took the form:

We vow never consciously to use our knowledge so as to bring harm to any of mankind, but rather to try to direct it so that the world may become a better place in which to live.

We liked this, were excited over it, and really meant it. Al produced a candle, which he lit as the sole light in the room. And, "sophomorically," by candlelight and with a triple handclasp, we made the pledge. We then discussed some of the ramifications of the pledge, and promised to renew it together periodically.

Contrary to what one may think, I also had friends who were not interested in mathematics. Among these was a delightful chap who undertook rambles with me. I purposely use the word "ramble" in contrast to "hike." What we did was an easy-going thing, with never more than 15 pounds on our backs; we scorned the hiker's 30-pound pack and his 20 miles a day.

Toward the end of the school year my friend's father was transferred to Ohio, not far from where the father's parents lived on a farm, and so the family had to move. My friend and I had a sober leave-taking, and we promised to write to one another. When the summer vacation came, I received a letter from my friend saying that he had secured the permission of his grandfather to invite me to spend two weeks on the grandfather's farm. Would I come? The rambling was excellent. I accepted the invitation, and took the long train ride to Ohio. It was a pleasurable journey. We snaked up the Alleghenies with two additional engines attached in tandem to our original engine. As one looked across the hairpin turns from the windows of the coach, you could see the three engines puffing along in the opposite direction to our own progress. At the crest of the range the additional engines were uncoupled and we proceeded across Pennsylvania, through Pittsburgh, across the needle of West Virginia, and into the rolling agricultural country of Ohio to the town of Wooster. There I was met by my friend and his grandfather, and in a horse-drawn carriage we drove north some eight or nine miles to the farm. By comparison with the huge western farms of today, the grandfather's farm was not very large, but it seemed immense to me. We enjoyed remarkable dinners around a great table in the evenings, and during the days we scrambled all over the farm, spending the hot afternoons in a cool swimming hole formed by a place where the creek that ran through the farm made a sharp right-angle turn.

Boys from neighboring farms came and swam with us. We naturally got to talking about school work, and several of the fellows produced notebooks that they had. These notebooks were of all sizes and shapes—some that were theme-paper in size and opened like books, some that were pads much like a stenographer uses, and some were tiny books in which one could only list homework assignments. The remarkable thing about all these notebooks was that in each of them, on a flyleaf and in beautiful Gothic lettering, appeared *The Scholar's Creed* by Dr. John J. Seelman. I'll never forget how impressed I was when I read that creed. Here is the main part of it, trimmed of a short preliminary and a short concluding paragraph.

> I believe the knowledge I have received or may receive from teacher and book, does not belong to me; that it is committed to me only in trust; that it still belongs and always will belong to the humanity which produced it through all the generations.
>
> I believe I have no right to administer this trust in any manner whatsoever that may result in injury to mankind, its beneficiary, on the contrary—
>
> I believe it is my duty to administer it singly for the good of this beneficiary, to the end that the world may become a kindlier, a happier, a better place in which to live.

Note that this creed contains the two points we made in the vow we took back in New Jersey, and that even the concluding words are exactly like the concluding words of our vow. But there is, in addition, a third point that we did not list, namely that knowledge is only on loan to us. It is claimed that some American Indians held this belief in connection with land. You do not own the land on which you live; it is only entrusted to you to use wisely and to preserve for those who come after you. When Rachel Carson finally made enough money, from royalties on her book *The Sea Around Us*, to purchase a headland of Maine, a friend asked her, "How does it feel to own a piece of Maine?" She replied, "Oh, I don't own it. It is only on loan to me to take care of for a while." In the same way that the American Indian claimed that we do not own the land we live on, *The Scholar's Creed* claims that we do not own the knowledge we possess. I could scarcely wait to return to New Jersey to report *The Scholar's Creed* to my two friends when we should again hold one of our little gatherings.

Al and Lou were duly impressed by *The Scholar's Creed.* It said so much more fully and so much better what we had tried to express that we decided, on the spot, to adopt it in place of our original vow.

The Perfect Game of Solitaire*

A list of some of the principal requirements for a good game of solitaire would surely include:

I. The rules of the game should be few and simple.

II. The game should not require highly specialized equipment, so that it can be played almost anywhere and at almost any time.

III. It should be truly challenging.

IV. It should possess a number of interesting variations.

Judged by the above requirements, the Greeks of over 2000 years ago devised what can perhaps be considered a perfect game of solitaire—it might now be called the *game of Euclidean constructions*.

The rules of the game are given in the first three postulates of Euclid's *Elements*. These postulates read:

1. A straight line can be drawn from any point to any point.
2. A finite straight line can be produced continuously in a straight line.
3. A circle may be described with any center and distance.

---

* For a considerably fuller treatment of this game see Chapter 3 of: Howard Eves, *Fundamentals of Modern Elementary Geometry*. Boston: Jones and Bartlett Publishers, 1992.

These postulates are the primitive constructions from which all other constructions in the *Elements* are to be compounded. Since they restrict constructions to only those that can be made in a permissible way with straightedge and compass, these two instruments, so limited, are known as the *Euclidean tools*.

The first two postulates tell us what we can do with a Euclidean straightedge; we are permitted to draw as much as may be desired of the straight line determined by any two given points. In other words, the length of the straightedge is not limited. The third postulate tells us what we can do with the Euclidean compass; we are permitted to draw the circle of given center and having any straight line segment radiating from that center as a radius—that is, we are permitted to draw the circle of given center and passing through a given point. Note that neither instrument is to be used for transferring distances. This means that the straightedge cannot be marked, and the compass must be regarded as having the characteristic that if either leg is lifted from the paper, the instrument immediately collapses. For this reason, a Euclidean compass is often referred to as a *collapsing compass*; it differs from a *modern compass*, which retains its opening and hence can be used as a divider for transferring distances.

It would seem that a modern compass might be more powerful than a collapsing compass. Curiously enough, such turns out not to be the case; any construction performable with a modern compass can also be carried out (in perhaps a longer way) by means of the collapsing compass. It is interesting that this fact is established by the first three propositions of Euclid's *Elements*. It follows that we may dispense with the Euclidean, or collapsing, compass. We are assured that the set of all constructions performable with straightedge and Euclidean compass is the same as the set performable with straightedge and modern compass. As a matter of fact, in all construction work, we need not be interested in actually and exactly carrying out the construction, but merely in assuring ourselves that such constructions are possible. To use a phrase of Jacob Steiner, we can do our constructions "simply by means of the tongue," rather than with actual instruments on paper. We seek then, the construction easiest to describe rather than the simplest or best construction actually to carry out with the instruments.

So much for the rules and equipment of the game. We come now to the challenging nature of the game. If one were asked to find the midpoint of a given line segment using only the straightedge, one would be justified in exclaiming that surely the Euclidean straightedge alone will not suffice, and that some additional tool or permission must be furnished. The same is true of the combined Euclidean tools; there are constructions that cannot be performed with these tools alone, at least under the restrictions imposed upon them. Three famous problems of this sort, which originated in ancient Greece, are:

1. *The duplication of the cube,* or the problem of constructing the edge of a cube having twice the volume of a given cube.
2. *The trisection of an angle,* or the problem of dividing a given arbitrary angle into three equal parts.
3. *The quadrature of the circle,* or the problem of constructing a square having an area equal to that of a given circle.

The fact that there are constructions beyond the Euclidean tools adds a challenge to the construction game. It becomes desirable to obtain a criterion for determining whether a required construction is or is not within the power of our tools.

But, in spite of the limited power of our instruments, really intricate constructions can be accomplished with them. Thus, though with our instruments we cannot, for example, solve the seemingly simple problem of drawing the two lines trisecting an angle of 60°, we can draw all the circles that touch three given circles (the *problem of Apollonius*); we can draw three circles in the angles of a given triangle such that each circle touches the other two and also the two sides of the angle (the *problem of Malfatti*); we can inscribe in a given circle a triangle whose sides, produced if necessary, pass through three given points (the *Castillon-Cramer problem*).

As for the fourth of the above listed requirements for a good game of solitaire, surely few games are as rich as the construction game. It would take much too much space to give a detailed account here. We will content ourselves by briefly mentioning only a few of the interesting variations of the game. The two principal variations are found in the following two remarkable theorems.

The Mohr-Mascheroni Construction Theorem. *Any Euclidean construction, insofar as the given and required elements are points, may be accomplished with the Euclidean compass alone*

The Poncelet-Steiner Construction Theorem. *Any Euclidean construction, insofar as the given and required elements are points, may be accomplished with the straightedge alone in the presence of a given circle and its center.*

In the Poncelet-Steiner construction theorem, the phrase "in the presence of a given circle and its center" may be replaced by "in the presence of two intersecting, tangent, or concentric circles," or by "in the presence of three noncoaxial circles," or by "in the presence of a given circle and a parallelogram." There are other interesting replacements of the phrase. Perhaps the most remarkable finding in connection with the Poncelet-Steiner theorem is that not all of the given circle is needed, but that *all Euclidean constructions are solvable with straightedge alone in the presence of a circular arc, no matter how small, and its center.*

August Adler and others have shown that *all Euclidean constructions are solvable with a double-edged ruler, be the two edges parallel or not*. Examples of the latter type of two-edged ruler are a carpenter's square and a draughtsman's triangle. It is interesting that while an increase in the number of compasses will not enable us to solve anything more than the Euclidean constructions, two carpenter's squares make it possible for us to duplicate a given cube and to trisect an arbitrary angle. These latter problems are also solvable with compass and a *marked* straightedge—that is, a straightedge bearing two marks along its edge. Various tools, such as a three-legged compass, a tomahawk, and linkages have been invented that will solve certain problems beyond those solvable by Euclidean tools alone. Another interesting discovery is that *all Euclidean constructions, insofar as the given and required elements are points, can be solved without any tools whatever, by simply folding and creasing the paper representing the plane of construction*. A further interesting result is that in making a Euclidean construction, the straightedge, which is assumed by Euclid to be as long as we might wish can be replaced by a straightedge of small finite length ε. The problem of the "best" Euclidean solution for a required construction has also been considered and a science of *geometrography* was developed in 1907 by Émile Lemoine for quantitatively comparing one construction with another. And, of course, the construction game has been extended to three-dimensional space. Basic in all construction work are such strategies as the method of loci, the method of transformation, and the method of coaxial homographic ranges.

Over the years I have played the construction game, and its variations, with great zest and assiduity, receiving immeasurable pleasure from the pursuit. I've played the game at all odd times—while waiting for a meeting to start, while rambling, in bed before falling asleep, and so on, and have developed a high skill at it. I have never wearied of playing the game. I can think of no other game, played alone, that can in any sense be compared to it. Chess, with its strategies, gambits, and variations, is a remarkable game for two players. But it involves to me the unpleasant feature of trying to beat someone; I do not find enjoyment in beating an opponent. Perhaps the construction game can be thought of as a sort of solitaire chess, free of that unpleasant feature. At any rate, it is an ideal game for someone, like myself, who tends to think in terms of pictures rather than in terms of words.

I should perhaps conclude by pointing out that Euclid used constructions in the sense of existence theorems—to prove that certain entities actually exist. Thus one may define a *bisector* of a given angle as a line in the plane of the angle, passing through the vertex of the angle, and such that it divides the angle into two equal angles. But a definition does not establish the existence of the thing being defined; this requires proof. To show that a given angle does possess a bisector, we show that this entity can actually be constructed. Existence theorems are very important in mathematics, and actual construction of the entity is the most satisfying way of proving its existence. One might define a *square circle* as a figure that is both a square and a circle, but one would never be able to prove that such an entity exists; the class of square circles is a class without members. In mathematics it is nice to know that the set of entities satisfying a certain definition is not just the empty set.

The Most Seductive Book Ever Written

Occasionally, when a group of people get together enjoying one another's company, someone will bring up the old hypothetical situation of a lone castaway stranded on a desert island. The island is "friendly," in the sense that it possesses a mild equitable climate and a plentiful supply of easily obtained food. Since the island is far from all sea lanes and air routes, the stranded individual knows he will remain there alone, unrescued, for a considerable time, perhaps years. Now if this individual can wish to have any book he desires washed ashore in a waterproof container, what book should he choose to help him while away his long lonely stay on the island?

Various answers have been given to the above question. At one gathering, an admirer of Elizabethan drama voted for a book containing the complete works of William Shakespeare; a good deal of time could be spent in a leisurely way enjoying the great masterpieces of this writer. A theologically inclined individual voted for an unexpurgated copy of the Christian Bible; he would have an opportunity to study this important work verse by verse. An armchair traveler voted for a complete atlas of the world; during his stay on the island this book would enable him to visit, in his imagination, every corner of the earth. There were a number of other worthy selections.

But to my mind there is a book that, for the purpose under consideration, vastly surpasses all other choices. That book is

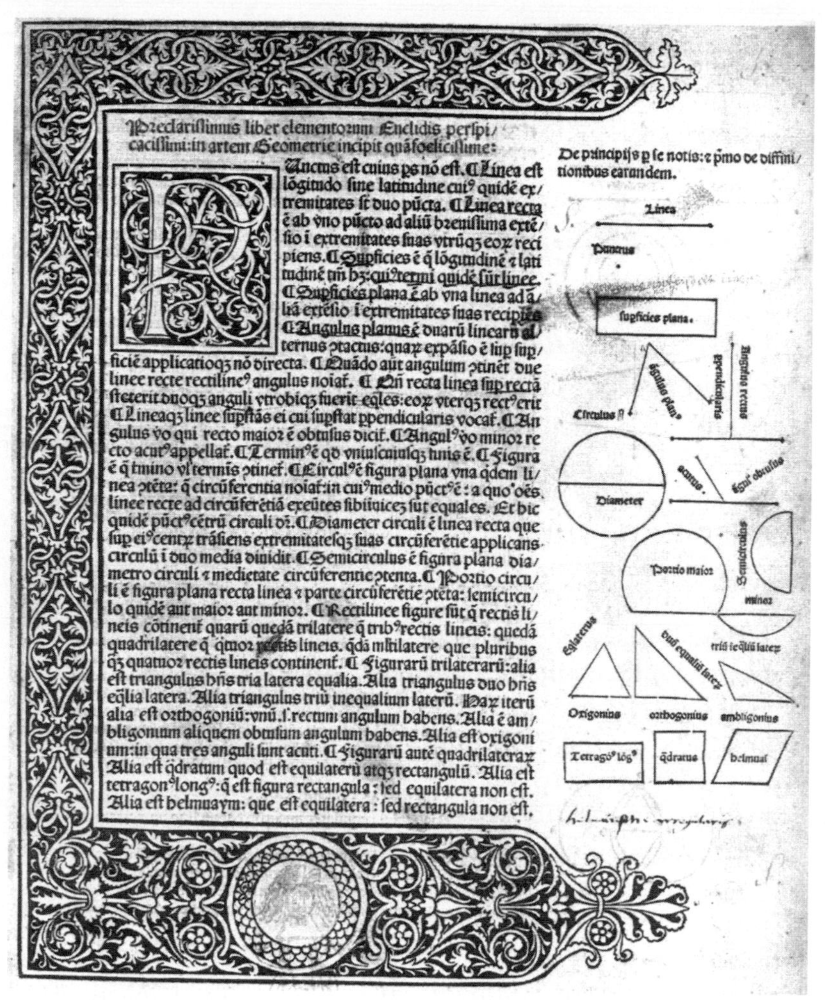

Preclarissimus liber elementorum Euclidis perspi/
cacissimi: in artem Geometrie incipit quā foelicissime:

Punctus est cuius ps nō est. ¶ Linea est
lōgitudo sine latitudine cuiꝰ quidē ex/
tremitates sūt duo pūcta. ¶ Linea recta
ē ab vno pūcto ad aliū breuissima extē/
sio i extremitates suas vtrūqꝫ eorꝝ reci
piens. ¶ Supficies ē q̄ lōgitudinē ꝛ lati
tudinē tm̄ hꝫ: cuiꝰ termi quidē sūt linee.
¶ Supficies plana ē ab vna linea ad a/
liā extēsio i extremitates suas recipiēs
¶ Angulus planus ē duarū linearū al/
ternus ꝯtactus: quarꝝ expāsio ē sup sup/
ficiē applicatioqꝫ nō directa. ¶ Quādo aūt angulum ꝯtinēt due
linee recte rectilineꝰ angulus noiaẗ. ¶ Qn̄ recta linea sup rectā
steterit duoqꝫ anguli vtrobiqꝫ fuerit eq̄les: eorꝝ vterqꝫ rectꝰ erit
¶ Lineaqꝫ linee supstās ei cui supstat ppendicularis vocaẗ. ¶ An
gulus vo qui recto maior ē obtusus dicit. ¶ Angulꝰ vo minor re
cto acutꝰ appellaẗ. ¶ Terminꝰ ē qd vniuscuiusqꝫ finis ē. ¶ Figura
ē q̄ tmino vl termis ꝯtineẗ. ¶ Circulꝰ ē figura plana vna q̄dem li/
nea ꝯtēta: q̄ circūferentia noiaẗ: in cuiꝰ medio pūctꝰ ē: a quo oēs
linee recte ad circūferētiā exeūtes sibiīuicēꝫ sūt equales. Et hic
quidē pūctꝰ cētrū circuli dr̄. ¶ Diameter circuli ē linea recta que
sup eiꝰ centꝝ trāsiens extremitatesqꝫ suas circūferētie applicans
circulū i duo media diuidit. ¶ Semicirculus ē figura plana dia/
metro circuli ꝛ medietate circūferentie ꝯtenta. ¶ Portio circu/
li ē figura plana recta linea ꝛ parte circūferētie ꝯtēta: semicircu/
lo quidē aut maior aut minor. ¶ Rectilinee figure sūt q̄ rectis li/
neis cōtineẗ quarū quedā trilatere q̄ tribꝰ rectis lineis: quedā
quadrilatere q̄ q̄tuor rectis lineis. q̄dā mltilatere que pluribus
qꝫ quatuor rectis lineis continent. ¶ Figurarū trilaterarū: alia
est triangulus hn̄s tria latera equalia. Alia triangulus duo hn̄s
eq̄lia latera. Alia triangulus triū inequalium laterū. Harꝝ iterū
alia est orthogoniū: vnū .s. rectum angulum habens. Alia ē am/
bligonium aliquem obtusum angulum habens. Alia est oxigoni
um: in qua tres anguli sunt acuti. ¶ Figurarū autē quadrilaterarꝝ
Alia est q̄dratum quod est equilaterū atqꝫ rectangulū. Alia est
tetragonꝰ longꝰ: q̄ est figura rectangula: sed equilatera non est.
Alia est helmuaym: que est equilatera: sed rectangula non est.

De principijs p se notis: ꝛ pmo de diffini/
tionibus earundem.

The first page of the first printed Euclid, printed by Ratdolt in Venice in 1482

any good British edition of the complete thirteen books of Euclid's *Elements*. I stipulate that it be a good British edition because those editions are particularly noted for their inclusion of copious and excellent allied problems. Let me list a few reasons for my choice.

1. Euclid's *Elements* can be read, without prerequisites, by essentially any one.

2. The work contains a total of 465 propositions covering fascinating properties of both plane and solid geometry and of the whole numbers—enough material to occupy one for a considerable time—all beautifully and charmingly deduced from a small handful of readily accepted axioms and postulates.
3. This book can occupy not only those hours spent studying the work itself, but also many other delightful hours of the day and night mulling over the fine problems and intriguing constructions posed in the book.
4. A pointed stick on the island beach would eliminate the need for pencil and paper, and allow erasures and corrections to be easily made.
5. In particular, the beach and pointed stick would furnish ideal equipment for playing the grand solitaire construction game.
6. The fine logic and beautiful unfolding of the material could excellently distract a castaway from his loneliness on the desert island.

It could well be that in such pleasant surroundings with this book for company, and with the complete lack of the usual interruptions and distractions of the world, the castaway would regard the appearance of any rescuing party with a measure of regret and disappointment.

For an illustration of the fine distracting powers of Euclid's *Elements*, consider the story told about Bernhard Bolzano when he once was suffering from an illness manifested by bodily aches and chills. To take his mind off his troubles, he picked up Euclid's *Elements* and for the first time read the masterly exposition of the Eudoxian doctrine of ratio and proportion set out in Book V. Lo and behold, his pains vanished. It has been said that thereafter, when anyone became similarly discomforted, Bolzano would recommend that the ill one read Euclid's Book V.

The Master Geometer

During my second year in the Graduate School at Harvard University I had both the honor and pleasure of working as a geometry research student under the great geometer Julian Lowell Coolidge, who, at that time, was probably the foremost geometer on the American continent. In addition to having a head crammed with an incredible stock of geometrical knowledge, Professor Coolidge possessed a charming wit

Julian Lowell Coolidge, 1925

and sense of humor. He once remarked to me, "When I teach I try to make my students laugh, and when their mouths are open, I put something in for them to chew on."

Julian Lowell Coolidge (1873–1954) was a descendent of Thomas Jefferson, a cousin of Abbott Lawrence Lowell, the mathematically inclined president of Harvard from 1908 to 1933. He was a Rough Rider under Theodore Roosevelt in the war of 1898. He studied in Europe under the eminent mathematicians Kowalewski, Segre, and Study from 1903 to 1904. He was a one-time teacher of Franklin Delano Roosevelt, at the famous Groton School for boys (founded in 1884). In 1925 he was elected president of the Mathematical Association of America and founded the Association's Chauvenet Prize in that year. He was appointed the first master of Lowell House at Harvard in 1930, and over the years published a masterful series of mathematics books, all bearing the Oxford University Press imprint.

Sandy

A relaxed atmosphere pervaded the monthly meetings of the Harvard Graduate Mathematics Club. Professor Coolidge often brought with him to the meetings his large loveable Airedale dog Sandy. Sandy would habitually choose a position on the rug directly in front of the middle of the blackboard. We who presented talks at the meetings were granted free use of the two ends of the blackboard, but the center piece was preempted by Sandy.

Conic
HYPERBOLA
ELLIPSE

The Perfect Parabola

Perhaps the most frequently told anecdote about Professor Coolidge occurred one day during his analytic geometry class, while discussing the conic sections. Coolidge had the habit of twirling his gold watch and chain, back and forth about his index finger as he lectured. At the start of a particularly vigorous swing, the chain broke, and the valuable watch looped across the classroom and landed, shattered on a stone window sill. Unperturbed Professor Coolidge immediately seized the lesson involved and uttered, "Gentlemen, you have just observed a perfect parabola."

Three Coolidge Remarks

1

Professor Coolidge's lectures were often enlivened with wit and humor. One day, in explaining the concept of a function passing to a limit, he stated, "The logarithmic function approaches infinity with the argument, but very reluctantly.

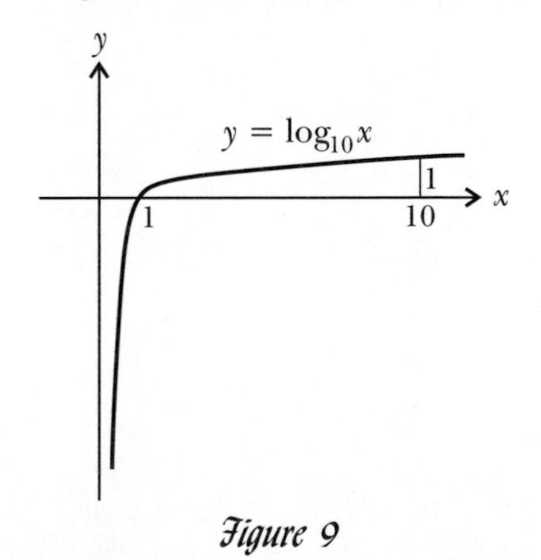

Figure 9

2

Professor Coolidge had a penchant for probability theory. In an early 1909 talk, entitled "The Gambler's Ruin," which concerned the effect of finite stakes on the prospects of a gambler, he proved that the best strategy is to bet the entire stake available on the first turn of a fair coin. “It is true,” he concluded, “that a man who does this is a fool. I have only proved that a man who does anything else is an even bigger fool.”

3

One day Professor Coolidge remarked to me that a nice thing about mathematics is that it never solves a problem without creating new ones.

Professor Coolidge During Examinations

Professor Coolidge used to pace around the classroom when his students were taking an examination. As he walked he would glance at the various students' efforts. If his eye fell on something that displeased him he would point it out to the student and advise him to rework the problem.

Professor Coolidge's Test

I recall how, at a large mathematical gathering, Professor Coolidge rose, advanced to the front of the room, and there frightened the group by announcing that he was going to give them a little mathematics test. Now mathematics professors may like to give tests, but to take one is quite another matter. To calm his audience, Professor Coolidge said he merely wanted to verify that most mathematicians know very little elementary solid geometry.

Professor Coolidge started by reviewing a few definitions, such as those of the *medians* and the *altitudes* of triangles and tetrahedra. "Now," he said, "though, as a high school student of geometry knows, the medians of a triangle are concurrent, can the same be said of the medians of a tetrahedron?" After some hesitation, almost everyone present said that surely they must be. Professor Coolidge assured them that this is indeed the case. He next similarly asked, "Though, as any high school student of geometry knows, the altitudes of a triangle are concurrent, can the same be said of the altitudes of a tetrahedron?" Many present said that of course they are concurrent, most of the others said that they blame well ought to be, and the few remaining ones, fearing some sort of a trap, were noncommittal. Professor Coolidge then explained that the altitudes of a tetrahedron usually are not concurrent, and that concurrency occurs only in the so-called *orthocentric tetrahedra*, in which each edge of the tetrahedron is perpendicular in space to the opposite edge of the tetrahedron.

Professor Coolidge went on to other simple questions and well established his thesis that indeed most mathematicians know very little elementary solid geometry.

Borrowing Lecture Techniques from Admired Professors

I dare say that most of us who lecture in college tend to incorporate in our style some of the techniques we admired in our own former professors. I know I have done this. The two professors I most admired at Harvard were David Vernon Widder and Julian Lowell Coolidge, and I have amalgamated lecture techniques of these two professors into my own lecturing.

Professor Widder was an absolute perfectionist. His lectures were meticulously prepared and delivered with great care. It

David Vernon Widder

was pure pleasure listening to him and following his inexorable progress through a topic. He was one of those lecturers a student wouldn't dream of interrupting, any more than one would interrupt Isaac Stern during one of his violin recitals. After all the students were quietly seated in his lecture room in Sever Hall, Professor Widder would emerge from a small door in the center of the front blackboard, launch into his impeccable lecture, and at the conclusion vanish through the little door. All questions were fielded by his very able assistant.

In contrast to Professor Widder were the enchanting, entertaining, and relaxed lectures of Professor Coolidge. Professor Coolidge was a great wit, and at times he was almost a clown during his presentations, exhibiting antics completely lacking in Professor Widder's very sedate lectures. There would be much laughter, and student questions eagerly sought.

It is the combined styles of these two professors that I have most incorporated into my own lectures. I try to give meticulously prepared presentations, but also recognize the value of occasional clownery.

My Teaching Assistant Appointment

Although, when I was an undergraduate at the University of Virginia, I was occasionally invited to assist Professors Luck and Linfield, it wasn't until my second year at Harvard that I became a genuine TA. The appointment was to assist Professor Ralph Beatley in one of his mathematics educa-

Prof. Ben Zion Linfield of the Mathematics Department of the University of Virginia

tion courses. Professor Beatley was the authority at Harvard in the field of mathematics education. Not only did I grade the quizzes and daily homework assignments of his class, but I pinch-hitted for him when he was away on speaking engagements. This occurred quite frequently, for Professor Beatley was a much sought-after speaker. Since Professor Beatley also taught the same course at nearby Radcliffe College for women, I also occasionally lectured there in his place. Professor Beatley was very kind to me, and later, when I was studying at Princeton, he took the time to call on me there during one of his speaking visits to Princeton.

A Night in the Widener Memorial Library

It's pretty certain that no one would be able to guess how I was introduced to the fascinating study of linkage machines. Here is the story.

We graduate mathematics students at Harvard were given keys to the extensive mathematics library housed in Room Q on the second floor of the university's Widener Memorial Library. Room Q was a large square room lined with well-stocked bookcases, and furnished with a number of very comfortable reading chairs and some small writing desks.

I spent a lot of time in Room Q, perusing many of the interesting books. I recall the delight I got paging through the extensive collected works of Arthur Cayley and the less extensive collected works of J. J. Sylvester. It was in Room Q that I encountered many great mathematics books for the first time.

One evening in Room Q, after checking some references for a piece of research I had been doing, I came across a slender little book with the alluring title *How to Draw a Straight Line; a Lecture on Linkages* (New York: The Macmillan Company, 1887)*, written by A. B. Kempe, a British barrister at law.

---

* This book has been reprinted in *Squaring the Circle, and Other Monographs*, Chelsea Publishing Company, 1953.

Widener Memorial Library

When I finished reading the little book, I looked at my watch and was surprised to find it was after 11 P.M., for the library closed each evening at 10 P.M. I put out the light and groped my way down stairs to the library entrance. I could see that the doors were locked, and fearing to tamper with them lest I should set off some loud burglar alarm, I regroped my way back to Room Q. There I settled myself in the dark in one of the comfortable reading chairs and spent a fairly comfortable night dozing and dreaming of enchanting linkage machines. In the morning, when I was sure the library was again open, I proceeded downstairs. On the way I encountered a janitor with whom I had become friendly, and I told him of my singular adventure. "So that was you in Room Q at closing time," he said. "Going down the corridor last night I saw the light through the transom above the door and yelled like a loon, 'Closing time.' Didn't you hear me?" No, I hadn't. A. B. Kempe apparently had closed my ears to all disturbing sounds.

When I returned to my lodgings, I found my landlady quite deranged. She was a motherly type, and aware that I had been absent all night, had started to call the hospital and the police station. Good Miss Rice. I often think of her and her kindly care of me.

As narrated above in MMM, in time my museum contained a nice collection of blackboard models of linkage machines. The subject of linkages became quite fashionable among geometers, and many linkages were found for constructing special curves. In 1933, R. Kanayama published a bibliography of 306 titles of papers and works on linkage mechanisms written between 1631 and 1931. It has been shown (in *Scripta Mathematica*, v. 2, 1934, pp. 293–294) that this list is far from complete, and of course many additional papers have appeared since 1931.

The Slit in the Wall

On Harvard Square, back in the mid-1930s, there was a small thin eating-place frequented by four of us graduate mathematics students. We called it *The Slit in the Wall*, and one day a surprising mathematical event took place there—in our favorite booth at the very rear of the place.

We had each enjoyed a big bowl of delicious minestrone soup when I casually announced that I had discovered an "obvious" proof of a very teasing relationship in number theory that had been eluding us for some time. The word "obvious" caused a ripple of surprise among my friends, along with a feeling of considerable skepticism.

"No," I insisted, "my proof is *truly* obvious, because the relationship can immediately be seen by merely looking at a simple geometrical diagram."

These words only increased the curiosity and incredulity of my friends—for what could a simple geometrical diagram possibly have to do with a problem in number theory? I produced a pen and asked if someone had a piece of paper. A piece of paper was not forthcoming, as we all, according to custom, had left our books and school things back at Sever Hall while we went to lunch. It was then that we noticed the proprietress of the establishment standing at our table, amused by the commotion we had made. We were all very fond of her—she had long given us special treatment and affectionately referred to us as "my mathematicians." Every day she greeted us with, "How are my mathematicians today?"

One of the group explained to her that I had found an amazing mathematical proof, but we had no paper so that I could show it to them.

"Here," she said, "put it on this napkin." And she handed me a good linen napkin. Horrified, I said I couldn't ruin a good cloth napkin by drawing on it.

"But I want you to," she replied. "I have a reason."

So I reluctantly took the napkin and with my pen drew on it the simple geometrical figure that visually exposed the teasing number theory relation. My three amazed companions heartily congratulated me, and the smiling proprietress gently took away the napkin.

A couple of days later, when we appeared at our favorite booth in The Slit in the Wall, lo and behold, there neatly stretched and framed behind glass on the wall of our booth was the linen napkin bearing my geometrical figure. Our good lady came and beamed at us.

Time has erased from my memory just what the original problem was. Indeed, at this distance in time, I cannot even recall for sure the proper name of our little Slit in the Wall, though the name Gusty's keeps recurring to me. Many years later I visited Harvard, but could find no trace of our little Slit in the Wall on Harvard Square.

Over the years I have devised a number of so-called "proofs without words," and have been delighted to see other such proofs appearing, usually as fillers, in *Mathematics Magazine* and other mathematical publications. For one not familiar with "proofs without words," here are two well-known ones of the Pythagorean Theorem—see Figures 10 and 11.

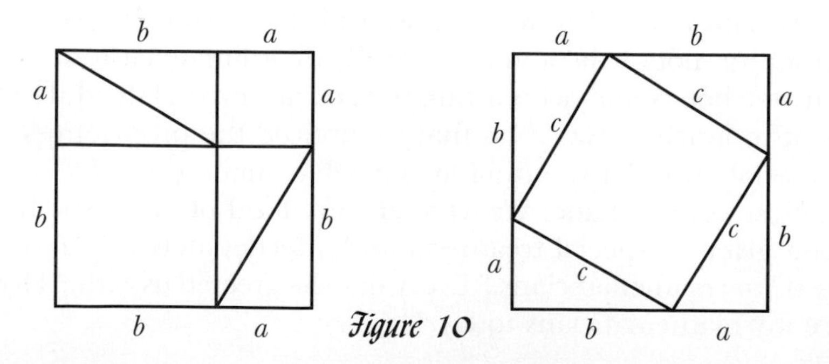

Figure 10

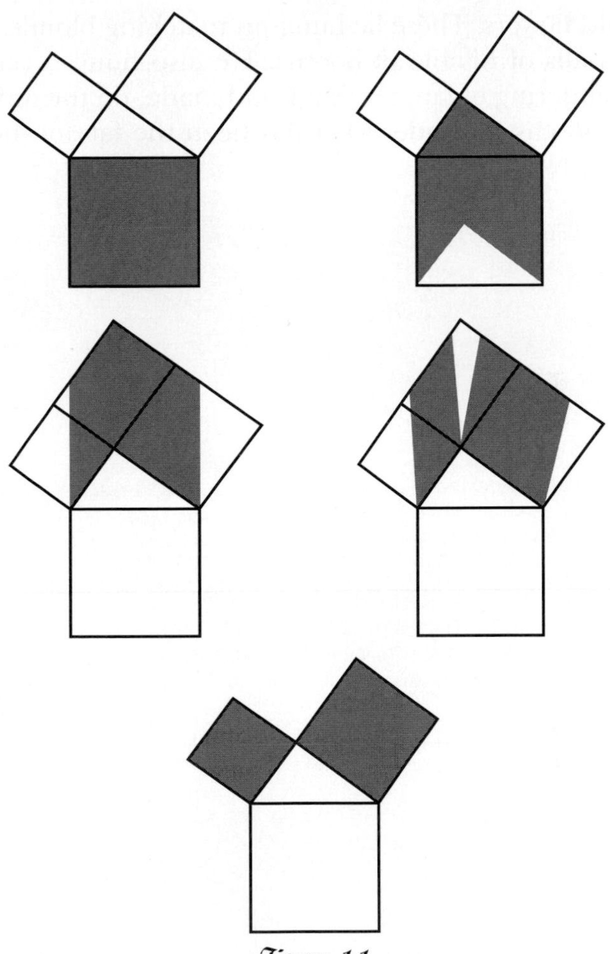

Figure 11

The idea of a proof without words is not a modern concept. As early as the twelfth century the Indian mathematician and astronomer Bhāskara gave a proof of the Pythagorean Theorem in which he sketched a figure and then simply appended to it the single word "Behold!"

Curiously, there was another Slit in the Wall that played a part in my life—back in the late 1940s. This slit was a long narrow dairy bar in Corvallis, Oregon. The proprietor of this slit commissioned me to paint about a dozen watercolors of Oregon wild flowers. These he hung, in matching blonde frames, on the walls of his file of booths. He also hung a very large pastel rendering of laurels, that I had made, on the wall opposite the booths. I wonder what has been the fate of these pictures.

Nathan Altshiller Court

I was an early, and then a continued, admirer of Nathan Altshiller Court. He wrote one of the first textbooks on college geometry. It was an excellent book and it did much toward popularizing that subject among college mathematics offerings. I corresponded a great deal with Professor Court, especially during my long tenure as editor of one of *The American Mathematical Monthly*'s Problems Sections, for he was an assiduous contributor to that department. But it wasn't until, upon a speaking tour, that I finally met him, at his university, the

Nathan Altshiller Court

University of Oklahoma. We had a delightful visit together, and during the visit I begged from him a copy of an attractive portrait of himself that I saw in his office. Smiling, he said he would gladly give me a copy, under the strict condition that I would hang it in my study overlooking my desk, where he could see that I kept busy at work. I agreed to the condition, and his portrait, now framed and behind glass, hangs in exactly such a position, overlooking my desk at Fox Hollow in Maine.

An Editorial Comment

In the manuscript for the fifth edition of my *An Introduction to the History of Mathematics*, I had written, in connection with the poorer Newtonian fluxional notation and the better Leibnizian differential notation, "The English mathematicians, though, clung long to the notation of their leader." The copy editor deleted the word "long," and wrote in the margin, "I thought Clung Long was a Chinese mathematician, not English."

THE ELEMENTS
OF GEOMETRIE
of the most aunci-
ent Philosopher
EVCLIDE
of Megara.

Faithfully (now first) translated into the Englishe toung, by H. Billingsley, *Citizen of London. Whereunto are annexed certaine Scholies, Annotations, and Inventions, of the best Mathematicians, both of time past, and in this our age.*

With a very fruitfull Præface made by M. I. Dee, *specifying the chiefe Mathematicall Sciēces, what they are, and wherunto commodious: where, also, are disclosed certaine new Secrets Mathematicall and Mechanicall, untill these our daies, greatly missed.*

Imprinted at London by *John Daye.*

The title page of the first Euclid to appear in English, translated by H. Billingsley and printed in 1570

Intimations of the Future

I've often been asked what it was that caused me to become interested in mathematics—and just when and how did it happen. This will be told in a later story. Here I will tell of a couple of foreshadowings of that wonderful event.

Some time in elementary school (I cannot now recall the precise grade) an excellent teacher introduced our class to the subject of geometrical areas. She first sketched on the board a rectangle 5 units long and 3 units wide. Then, drawing lines parallel to the sides of the rectangle through the unit divisions of those sides, she divided the rectangle into an array of small unit squares. How many of these squares were there? Well, there were 3 rows of 5 unit squares apiece, giving a total of $3 \times 5 = 15$ unit squares in all. Thus the area of the rectangle in unit squares is given by the product of its two dimensions. Clearly the same would be true of any other rectangle having integral dimensions. Though at this stage, she said, it would be too difficult to prove that in any case, whether the sides are integral or not, the area of a rectangle is given by the product of its two dimensions, we will assume that this is indeed the case.

Next she drew a parallelogram on the board as shown in Figure 11, and dropped perpendiculars from the two top opposite vertices. Clearly the parallelogram has the same area as the obtained rectangle, for each is made up of the central piece along with one of the two congruent right triangles. It follows that the area of the parallelogram pictured in Figure 11 is given by the product of its base and the altitude on that base. The

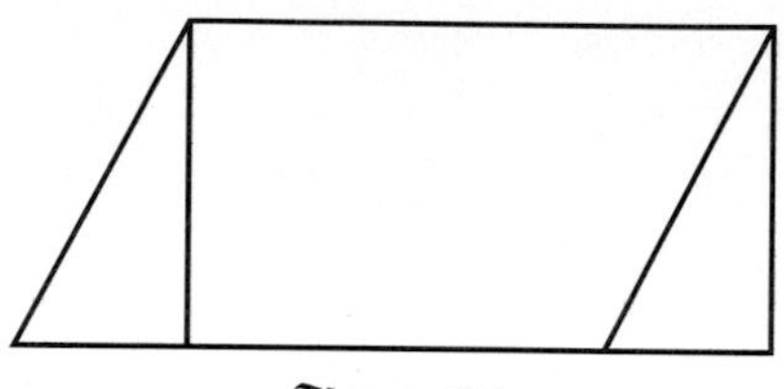

Figure 12

teacher failed to show that the reasoning from her figure does not necessarily apply to all parallelograms with a randomly chosen side as base. However, the exceptional situations are easily taken care of by slightly adjusting the reasoning, thus assuring that the area of a parallelogram is given by the product of *any* side of the parallelogram and the altitude on that side—a fact that is needed in further parts of the teacher's treatment of areas. (I later filled in the hiatus myself. See the concluding Addendum.)

She now drew a triangle on the board, and by attaching another one congruent to it, as shown in Figure 12, she obtained a parallelogram. Thus the area of a triangle is given by half the product of any side of the triangle and the altitude on that side. In particular, the area of a right triangle is then half the product of its two legs. By drawing the two diagonals of a rhombus, it is seen that the area of a rhombus is given by half the product of its two diagonals.

Next came the trapezoid. By drawing a diagonal of the trapezoid, thus dividing it into two triangles, one sees that the area of a trapezoid is given by half the product of the sum of its two bases and its altitude.

She now considered a regular polygon. By connecting each vertex of the polygon with the center of the polygon, the figure was divided into a number of congruent isosceles triangles, the area of each triangle being given by half the product of a side

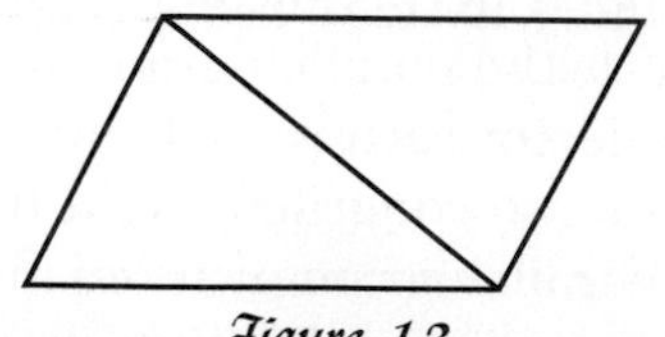

Figure 13

of the polygon and the apothem[1] of the polygon. Thus the area of a regular polygon is given by half the product of its perimeter and its apothem.

Finally she drew a circle, and inscribed in it a regular polygon of a large number of sides. By steadily doubling the number of sides of the inscribed polygon, one sees that the perimeter of the polygon approaches the circumference of the circle, and the area of the polygon approaches the area of the circle. It follows that the area of a circle is given by half the product of its circumference and its radius.

It was the neat step-by-step procedure that so fascinated me—not the individual formulas, but the inexorable progress starting from the readily assumed formula for the area of a rectangle.

The other mathematics of elementary school consisted of arithmetic, and held little charm for me. Nor was I enamored of the memorization of the addition and multiplication tables. My curiosity was aroused, however, when we were taught the algorithms for long multiplication and division. It seemed remarkable that they worked, and the teacher was unable to tell us why they worked. But it was the algorithm for finding square roots that seemed most mysterious to me. The process taught to us was the one where you start marking off the digits in the given number in pairs in both directions from the decimal point. By a marvelous procedure one found the sought square root, digit by digit.[2]

Again, the teacher was unable to justify the procedure. I spent many fruitless hours trying to figure out why the procedure worked. It wasn't until I took algebra in high school and met the identity

$$(a + b)^2 = a^2 + 2ab + b^2,$$

that the procedure finally fell apart.

[1] I recall that in the class the kids began to call the *apothem* the *apple stem*.

[2] This is not the method currently taught in the schools.

Addendum

From the figure

$$ABCD = AFD + GCD - BFG$$

and

$$FECD = BEC + GCD - BFG.$$

But $AFD \cong BEC$. Therefore $ABCD = FECD$.

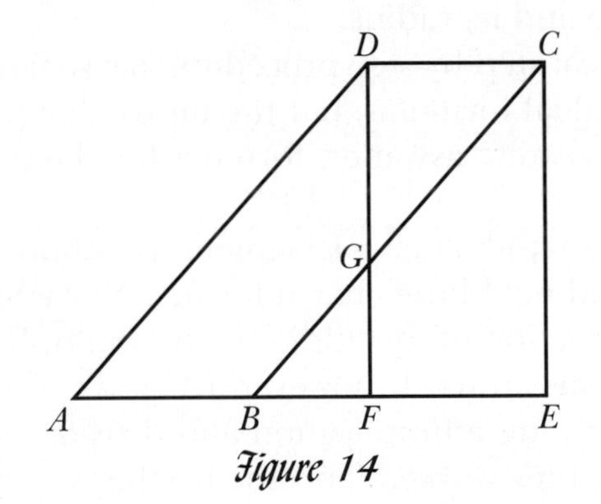

Figure 14

A Rival Field

Occasionally I am asked if there is some field I might have cared to go into instead of mathematics. Though there is a rival area that has strongly pulled on me, the conviction that I would have done so much more poorly in it than in mathematics makes the answer a definite "No." Not only that, but I had already become dedicated to mathematics. That other area is music. My desires there would have been too high for me to attain, for I would want to compose *outstanding* symphonies—that, or nothing at all. So I content myself to listening to great symphonies rather than trying in vain to write them.

There is certainly a strong kinship between music and mathematics. J. J. Sylvester had a keen sense of this. In a paper of his, "On Newton's rule for the discovery of imaginary roots," he exclaims in a footnote:

"May not Music be described as the Mathematic of sense, Mathematic as Music of the reason? the soul of each the same! Thus the musician *feels* Mathematic, the mathematician *thinks* Music,—Music the dream, Mathematic the working life—each to receive consummation from the other when human intelligence, elevated to its perfect type, shall shine forth glorified in some future Mozart-Dirichlet or Beethoven-Gauss—a union already indistinctly foreshadowed in the genius and labours of Helmholtz!"

During my Harvard years I had great opportunity to enjoy fine music, for just across the Charles River in Boston was the world-famous Boston Symphony Orchestra, conducted at that time by the incomparable Serge Koussevitzky. I subscribed to the Saturday night classical series of concerts, securing a seat in the very center of the first row of the balcony. How I enjoyed those evenings! I would go over to Boston early, dine in a pleasant and quiet little restaurant close to Symphony Hall, and then in a leisurely manner proceed to the Hall. But how I hated at the end of each performance to break the beautiful spell by training back to Cambridge. I think an ideal Symphony Hall should be provided with comfortable little sleeping booths to which a listener could retire for the night at the end of the concert, there to fall gently asleep while the orchestra continues softly to play some enchanting encore.

A Chinese Lesson

When I was teaching at Oregon State College (now Oregon State University), I designed and taught my first course in the history of mathematics. I entitled the course *Great Moments in Mathematics*. The title was motivated by a sequence of musically illustrated lectures, entitled Great Moments in Music, given some years ago on radio by the illustrious musicologist Walter Damrosch.

My course consisted of 60 separate "great moments," and each lecture was accompanied by an assortment of visual aids in the form of desk experiments, models, portraits, maps, overhead transparencies, and so forth. The course prospered very gratifyingly—to such an extent that I had to be moved into larger and larger lecture rooms.

In addition to regularly enrolled college students, many of the college's faculty attended. And whenever a scholar or group of scholars visited the college, the President of the college always advised them to attend one of my history lectures. It was in this way that, during an early lecture of the course devoted to various mathematical numeral systems, a small delegation of Chinese scholars, visiting many of the American colleges and universities along our west coast, passed through Corvallis and visited Oregon State College. Thus I was surprised one morning, shortly before starting my lecture, to see the delegation quietly enter and seat themselves in my lecture room.

I gave my lecture, illustrating the various numeral systems on an enlarging screen. When the lecture was over, the Chinese delegation came forward and, with much bowing, very fully and graciously congratulated me on my lecture. Then, one member of the group, with a show of embarrassment, asked if I would accept a small suggestion. Upon acknowledging that I would, he informed me that the brush strokes employed in forming the various Chinese numerals that I had illustrated, were always made in a definite order, and were so taught in all the Chinese elementary schools. I was delighted to learn these proper procedures, for I had been merely copying the various symbols in a more or less random fashion.

Notes. When, some years later, I accepted a position as chairman of the Mathematics Department of Champlain College (in Plattsburgh, New York), a newly formed unit of the State University of New York, I was urged by my good friend Carroll Newsom to write up my separate *Great Moments* lectures into a single consecutive story of the history of mathematics to be used in teaching the subject in colleges. Dr. Newsom was, at the time, among many other things, the mathematics editor of Rinehart Book Company. I acquiesced to Dr. Newsom's suggestion, and thus, in 1964, was born my *An Introduction to the History of Mathematics*. After 37 years, in a sixth edition and under a different publisher, that book is still prospering, widely used in many colleges in America and abroad. It has been translated into Spanish and Chinese, and has been called "The Rolls Royce of the textbook industry." In 1980, the subcommittee of the Dolciani Mathematical Expositions of the Mathematical Association of America asked me to submit a typescript of my original *Great Moments* for publication. I selected 40 of the 60 "great moments," which now appear in the two little volumes: *Great Moments in Mathematics Before 1650* and *Great Moments in Mathematics After 1650*.

The Bookbag

The mention, in an earlier item, that I have lectured in every state of the union, except Alaska, and indeed many times in some states, brings to my mind one of the most prized gifts I have ever received.

When I was a boy, my mother gave me an elegant bookbag, or brief case, made of sturdy artistically rippled black leather, equipped with three interior compartments, a strong handle, six little silver hemispheres on its bottom to protect it when it is set down, a little lock

In prep school (1930) at Chauncey Hall in Boston

with a tiny key, and my initials H.W.E., in bright silver letters, beautifully impressed on the bag's closing leather band.

I promised my mother that the bag would accompany me to all my school and college classes. As time went on, it has also accompanied me to all my lecture classes, and to all the many invited lectures that I have given. Though the bookbag is now somewhat travel-worn, it will continue to go with me wherever my academic life may lead me. I wish my mother could know the many hundreds of places the bookbag has accompanied me. When I come to the end of my academic career I feel the faithful old bookbag should be permanently bronzed.

Running the Mile in Twenty-one Seconds

I have an identical twin brother, Don, and over the years we have had much fun confusing people—especially our teachers when we were in grade school together. Don's field is biology, and for years he taught biology and general science in high schools and prep schools. He also possesses a quirk for harmless practical jokes.

One fall day I visited Don when he was teaching at the Mohonk Prep School on Lake Mohonk in New York State. The school was operated in English style, with masters and forms instead of teachers and grades. When I arrived at his school I went directly to his living quarters. He asked me if anyone had seen me, and I replied that I didn't think so. "Good," he said, "we'll have a little fun this afternoon after the last classes meet." He went on and said that in his up-coming general science class he would express to his kids his amazement that it took so long before the mile was run in less than four minutes, because he himself could do it in much less time than that. "Of course, the kids will not believe me," he said. "So I will arrange a demonstration at three o'clock on the trail that runs around the lake. You will hide in some bushes back down the trail, and at the shot of a cap pistol I will take off up the trail, which runs a mile around the lake. Shortly after I take off and disappear around a bend in the trail, you will run in from your hiding place." And so things were arranged. We both dressed in Don's customary

Lake Mohonk Mountain House from Thurston Rock, Lake Mohonk, NY

khaki trousers and shirts, and after his last class I positioned myself in a clump of bushes about fifty feet down from where Don was to set out on his race. I waited there and saw Don and the kids arrive.

Don put on quite an act—prancing to limber up for his dash, digging starting toe holes, and positioning himself for a rapid take-off. A kid shot off a cap pistol and Don set off up the trail, soon vanishing around a bend. The kids all stood staring in the direction of Don's departure. After a very short time I came pounding in from my hiding place. The astonished kids all turned around, eye-popping and open-mouthed, as I tore to the finish line. One kid said, "He isn't even out of breath." And the time-keeper gasped, "He did it in only twenty-one seconds." I'd never seen a group of such amazed kids. After several minutes, Don reappeared, and we exposed the hoax. The kids now began rubbing their eyes—from seeing double.

Winning the 1992 Pólya Award

Two planar pieces that can be placed so that they intercept chords of equal length on each member of some family of parallel lines, or two solid pieces that can be placed so that they intercept sections of equal area on each member of some family of parallel planes, are said to be *Cavalieri congruent*.

Some years ago I proved the following two theorems which, at first encounter scarcely seem to be true:

Though one can exhibit two tetrahedra of equal volume that are *not* Cavalieri congruent, *any two triangles of equal area are Cavalieri congruent*.

Though one can easily show that there exists *no* polygon Cavalieri congruent to a circle, *there exists a tetrahedron Cavalieri congruent to a sphere*.

It was my publishing of these two theorems that won me the 1992 Pólya Award. The certificate of the award was accompanied by the following paragraph.

> This short article really packs a wallop! The author's two theorems indeed "at first encounter scarcely seem to be true," and the proofs are excellent illustrations of the power and beauty of geometry. The Cavalieri equivalence of a sphere and a tetrahedron is truly memorable—the sort of result which geometers in ancient times would have inscribed on their tombstones. The article is graced with the author's historical scholarship and lucid prose.

Cavalieri equivalence of a sphere and a tetrahedron*

Theorem. *There exists a tetrahedron to which a given sphere is Cavalieri congruent.*

Proof. Denote the radius of the given sphere by r. In the planes tangent to the sphere at its north and south poles, draw (see Figure 15) two line segments AB and CD perpendicular to one another, each of length $2r\sqrt{\pi}$ and having the line segment joining their midpoints as a common perpendicular.

Form the tetrahedron $ABCD$. The equatorial plane of the sphere cuts the tetrahedron in a square of side $r\sqrt{\pi}$. Let a plane parallel to the equatorial plane and at a distance x from it cut the sphere in a circle and the tetrahedron in a rectangle of sides u and v, where u is parallel to AB and v is parallel to CD. From the figure we see that the circular section of the sphere has area $\pi(r^2 - x^2)$. We also see, from similar triangles, that

$$\frac{u}{r\sqrt{\pi}} = \frac{r+x}{r}, \qquad \frac{v}{r\sqrt{\pi}} = \frac{r-x}{r},$$

whence

$$uv = \pi(r + x)(r - x) = \pi(r^2 - x^2).$$

It follows that the sphere and the tetrahedron are Cavalieri congruent.

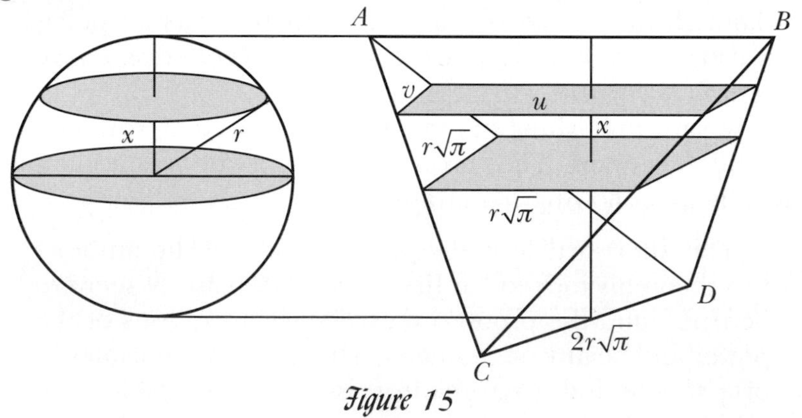

Figure 15

*Reprinted from *The College Mathematics Journal,* vol. 22, no. 2, March 1991, pp. 123–124.

A Love Story

When I was in the final grade of elementary school I fell madly in love. It was a clear case of love at first sight, and it evolved into an all-consuming passion of adoration that has continued during the entire rest of my life. I had fallen head over heels in love with the enticingly beautiful goddess Mathesis. Let me tell how it came about and then expand on some of its delightful joys.

I, and my two brothers, were brought up in Paterson, New Jersey. On occasional Saturday mornings the three of us would go down Broadway to a church that ran a couple of hours of free movies for children. After the movies we would often cross the street to the Danforth Memorial Library—usually in the hope of finding there a Sherlock Holmes book we hadn't yet read. Entering the library one Saturday morning I saw a short row of books displayed on the librarian's desk. They were various British editions of Euclid's *Elements*. I pulled out one of the books, entitled *The Harpur Euclid*, and discovered that it contained an enormous number of geometrical facts all presumably deduced from a small handful of very readily accepted initial assumptions laid down at the very start of the work. I was instantly amazed and awestruck, and asked the librarian if I could take the book out. Not that one, she said, since it was on display, but she could give me a copy from the stacks. I eagerly accepted the offer, and almost from that moment on I have lived in a state of high euphoria.

How the book engrossed me! It was devoted to the first six books of Euclid's *Elements*. I soon obtained a copy of my own,

THE FIRST SIX BOOKS OF

THE ELEMENTS OF EUCLID

IN WHICH COLOURED DIAGRAMS AND SYMBOLS

ARE USED INSTEAD OF LETTERS FOR THE

GREATER EASE OF LEARNERS

BY OLIVER BYRNE

SURVEYOR OF HER MAJESTY'S SETTLEMENTS IN THE FALKLAND ISLANDS

AND AUTHOR OF NUMEROUS MATHEMATICAL WORKS

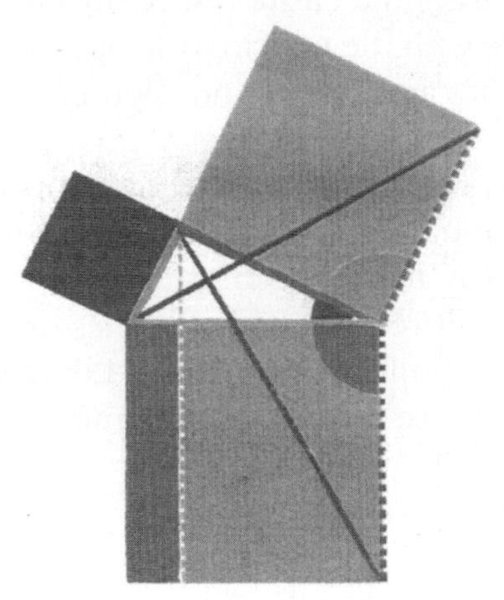

LONDON

WILLIAM PICKERING

1847

A curious edition of Euclid, printed in 1847, in which arguments are made entirely with colored diagrams—the work of Oliver Byrne, "Surveyor of Her Majesty's Settlements in the Falkland Islands"

and religiously and painstakingly went through it, proposition by proposition. It contained copious and attractive problems, or "riders" as they were called, and I worked on all of them, keeping my findings in a notebook. I found Book V very challenging and Book VI a veritable treasure trove of beautiful results. It wasn't long before I was convinced that I had in my hands one of the greatest books ever written.

It's no wonder that I came to feel that for some reason Mathesis had deliberately selected me as one of her lovers, and that she must love me in return. Somehow she guided me to buy my second significant mathematics book, Cajori's *A History of Mathematics*, so that I would see the future wonderful things she could offer me. I was never jealous that she loved many others besides myself—I was just overjoyed that she loved me. And how I loved her! In time she helped me climb a great peak from which the view of the surrounding country below was breathtaking—the peak of Functions of a Complex Variable. Hand in hand we walked through a beautiful flower-strewn meadow—the field of Abstract Algebra. We bathed together under a mountain waterfall—the Galois Theory of Equations. We stood side by side along a broad and beautiful flowing river—Felix Klein's Erlanger Programm. We stepped through Alice's mirror—into the land of Non-Euclidean Geometry. We slept in each other's arms on an ocean beach, lulled by the roar of the surf—the glistening sands of Differential Geometry. We loitered along a babbling brook—the stream of the Calculus of Variations. We walked among great redwood trees—the grove of Projective Geometry. We canoed on mirror-like waters—the lake of Mathematical Logic. We strolled beneath the stars on a clear and quiet night—the evening of the Philosophy of Mathematics. And we were together in many other memorable places.

As the years have gone by I have aged, but Mathesis has remained as young and beautiful as ever. It was some time back that I vowed I would try to share her beauty with others, and so have spent a large part of my life teaching in university classrooms, lyrically extolling her attractions. Perhaps I have managed to win over others to serve under the gracious goddess Mathesis.

Thank you, my sweet goddess, for the exquisite pleasure you have given me.

Addenda

1. A great many mathematicians found their beginnings in, or were markedly stimulated by, the enchanting material of Euclidean geometry.

In later years Einstein claimed that an enormous emotion was aroused in him when in his twelfth year an uncle gave him a small textbook on Euclidean geometry. The book utterly absorbed his interests, and later, in his *Autobiographical Notes,* he wrote with rapture of "the holy little geometry book." He says: "Here were assertions, as for example the intersection of the three altitudes of a triangle in one point, which—though by no means evident—could nevertheless be proved with such certainty that any doubt appeared to be out of the question. The lucidity and certainty made an indescribable impression on me." When Einstein passed away, "the holy little geometry book" was found still among his treasured possessions. It is now preserved in a filing cabinet at Princeton University.

Bertrand Russell began the study of Euclid's *Elements* when he was eleven, with his eighteen-year-old brother serving as tutor. In his *Autobiography*, Russell says: "This was one of the great events of my life, as dazzling as first love. I had not imagined there was anything so delicious in the world. ... From that moment until ... I was thirty-eight, mathematics was my chief interest and my chief source of happiness."

2. Unquestionably a high school student's gateway to mathematics is his course in Euclidean geometry, and not his course in introductory algebra. It is in the geometry course that he encounters (and in a strong dose) the essential mathematical ingredient of deduction. In contrast, high school algebra is largely a collection of procedures, with almost no deduction employed. It isn't until a student reaches abstract algebra in college that algebra becomes genuine mathematics.

3. A fair working rule for determining whether an incoming college student will do well in mathematics is to ask the student how he or she did in high school geometry and algebra. If the reply is: "I did quite well in algebra, but poorly in geometry," then that student probably *is not* a potential mathematics student. On the other hand, if the reply is: "I did real well in geometry, but algebra held much less interest for me," then that student probably *is* a potential mathematics student.

Eves' Photo Album

As a teenager in Hopatacong, NJ I was an enthusiastic sailor of small boats.

Playing with two of our children, Cindy and Jamie

Diane and I at her parents' home in Oviedo, FL

Time with granddaughter Courtney

Receiving my first honorary degree—Dr. of Science—
from the President of Western Maryland College (1990)

West Quoddy Light, the easternmost point in the U.S., a few miles from Lubec

A Condensed Biography of Howard Eves

In 1976 Professor Howard Eves retired from the University of Maine at Orono after a long and distinguished career as a teacher, geometer, writer, editor, and historian of mathematics. Since then he and his wife have divided their time between Lubec, Maine and Oviedo, Florida, with the choice appropriate to the season. From 1986 to 1991 he taught at the University of Central Florida in Orlando.

Howard Whitley Eves was born in Paterson, New Jersey on January 10, 1911, one of identical twin brothers. Following graduation from high school in Paterson he attended the University of Virginia, graduating with a BS in mathematics in 1934. After two years at Harvard, at which he earned a master's in mathematics, he moved to Princeton where, after a year, his studies were interrupted by the death of his father and the problems created by the Depression. He worked as a land surveyor in New Jersey, taught for a year at Bethany College in West Virginia and served as a mathematician for the Tennessee Valley Authority from 1937 to 1942.

After short stays at Syracuse University and the College of Puget Sound, he became an Associate Professor of Mathematics at Oregon State College (now University), where he received a Ph.D. in mathematics in 1948. After further appointments at Champlain College and Harpur College he moved to the University of Maine in 1954. He has lectured in every state of the

Union but one—Alaska—and has been active in various mathematical organizations. He has served as associate editor of the *American Mathematical Monthly*, *Mathematics Magazine*, the *Two-Year College Mathematics Journal*, the *Mathematics Teacher*, and the *Fibonacci Quarterly*. For 25 years he was Elementary Problems Editor of the *Monthly* and for several years edited the Historically Speaking section of the *Mathematics Teacher*. On the occasion of his 80th birthday, he was honored with an International Conference on Geometry, Mathematics History and Pedagogy, held at the University of Central Florida. For his writing and teaching he has won awards too numerous to list here, but they include a Distinguished Achievement Award from the State of Maine and the University of Maine, and honorary doctorates from the University of Maine and Western Maryland College. From the MAA he received the George Pólya Award for writing in 1992.

The last award is indeed fitting since by most people he is best known for his nearly twenty-five books, including what is probably the most widely used text in courses in the history of mathematics: *Introduction to the History of Mathematics*, now in its 6th edition. His *Mathematical Circles* books are much admired and treasured by his fans, consisting as they do of hundreds of charming mathematical stories, historical notes, and anecdotes. His two-volume *Survey of Geometry* is a classic, combining geometry with numerous asides to put the mathematics in its proper historical context. The list of his books goes on and on (see the abridged bibliography below). In addition he has published 150 articles in a variety of fields.

Eves' Lubec home is a 150-year old house—Fox Hollow—near the easternmost town in the United States, the last stop on the way to Campobello Island in New Brunswick, and with nearby views of the Bay of Fundy. It has been an ideal home for someone like Eves whose two favorite hobbies are hiking and watercolor painting. His passion for hiking resulted in one extraordinary six-month hike along the 2200 mile Appalachian Trail, from Mount Katahdin in Maine to Mount Oglethorpe in Georgia.

— Gerald L. Alexanderson, Editor

An Abridged Bibliography of Howard Eves' Works

Books

College Geometry, Jones & Bartlett, 1995.

Elementary Matrix Theory, Dover, 1980.

Foundations and Fundamental Concepts of Mathematics (3rd ed.), Dover, 1997. (Translation: Korean)

Functions of a Complex Variable (2 vol.), Prindle, Weber & Schmidt, 1966.

Fundamentals of Modern Elementary Geometry, Jones & Bartlett, 1992.

Great Moments in Mathematics (Before 1650), Mathematical Association of America, 1980.

Great Moments in Mathematics (After 1650), Mathematical Association of America, 1981.

In Eves' Circles (with Joby Milo Anthony), Mathematical Association of America, 1994.

Introduction to the History of Mathematics (6th ed.), Saunders, 1994. (Translations: Chinese, Spanish)

Mathematical Circles Series:

In Mathematical Circles (2 vol.), Prindle, Weber & Schmidt, 1969.

Mathematical Circles Revisited, Prindle, Weber & Schmidt, 1971.

Mathematical Circles Squared, Prindle, Weber & Schmidt, 1972.

Mathematical Circles Adieu, Prindle, Weber & Schmidt, 1977.

Return to Mathematical Circles, Prindle, Weber & Schmidt-Kent, 1987.

A Survey of Geometry (2 vol.), Allyn & Bacon, 1963, 1965. (Translation: Spanish)

The Other Side of the Equation, Prindle, Weber & Schmidt, 1971.

Translations

Initiation to Combinatorial Topology by Maurice Fréchet and Ky Fan, Prindle, Weber & Schmidt, 1967.

Introduction to the Geometry of Complex Numbers, by Roland Deaux, Ungar, 1957.

Chapters in Books

"Analytic Geometry," "Curves and Surfaces," "Mensuration Formulæ," and "Trigonometry," in *CRC Handbook of Tables for Mathematicians*

"Crises in the Foundations of Mathematics," in *Mathematics: Our Great Heritage*, by W. L. Schaaf, ed., Harper, 1948

"The History of Geometry," in *Historical Topics for the Mathematics Classroom, 31st Yearbook of the National Council of Teachers of Mathematics*, 1969

"Slicing It Thin," in *The Mathematical Gardner*, D. A. Klarner, ed., Prindle, Weber & Schmidt, 1981

Co-edited Books

The Otto Dunkel Memorial Problem Book (with E. P. Starke), Mathematical Association of America, 1957.

Over 150 articles in various journals and in the *Encyclopedia Americana, Collier's Encyclopedia,* and the *World Book Encyclopedia*.

Index